Statistics is Easy!

Second Edition

Synthesis Lectures on Mathematics and Statistics

Editor
Steven G. Krantz, *Washington University, St. Louis*

Statistics is Easy!, Second Edition
Dennis Shasha and MandaWilson
2010

Lectures on Financial Mathematics: Discrete Asset Pricing
Greg Anderson and Alec N. Kercheval
2010

Jordan Canonical Form: Theory and Practice
Steven H.Weintraub
2009

The Geometry of Walker Manifolds
Miguel Brozos-Vázquez, Eduardo García-Río, Peter Gilkey, Stana Nikcevic, and Rámon
Vázquez-Lorenzo
2009

An Introduction to Multivariable Mathematics
Leon Simon
2008

Jordan Canonical Form: Application to Differential Equations
Steven H.Weintraub
2008

Statistics is Easy!
Dennis Shasha and MandaWilson
2008

A Gyrovector Space Approach to Hyperbolic Geometry
Abraham Albert Ungar
2008

Statistics is Easy!, Second Edition
Dennis Shasha and Manda Wilson

ISBN: 978-3-031-01272-3   paperback

ISBN: 978-3-031-02400-9   ebook

DOI: 10.1007/978-3-031-02400-9

A Publication in the Springer' series

SYNTHESIS LECTURES ON MATHEMATICS AND STATISTICS

Lecture #8

Series Editor: Steven G. Krantz, Washington University St. Louis

Series ISSN
Synthesis Lectures on Mathematics and Statistics
Print: 1938-1743   Electronic: 1938-1751

10 9 8 7 6 5 4 3 2 1

Statistics is Easy!, Second Edition

Dennis Shasha
Department of Computer Science
Courant Institute of Mathematical Sciences
New York University

Manda Wilson
Bioinformatics Core
Computational Biology Center
Memorial Sloan-Kettering Cancer Center

SYNTHESIS LECTURES IN MATHEMATICS AND STATISTICS #8

ABSTRACT

Statistics is the activity of inferring results about a population given a sample. Historically, statistics books assume an underlying distribution to the data (typically, the normal distribution) and derive results under that assumption. Unfortunately, in real life, one cannot normally be sure of the underlying distribution. For that reason, this book presents a distribution-independent approach to statistics based on a simple computational counting idea called resampling.

This book explains the basic concepts of resampling, then systematically presents the standard statistical measures along with programs (in the language Python) to calculate them using resampling, and finally illustrates the use of the measures and programs in a case study. The text uses junior high school algebra and many examples to explain the concepts. The ideal reader has mastered at least elementary mathematics, likes to think procedurally, and is comfortable with computers.

All code was tested using Python 2.3.

Note: When clicking on the link to download the individual code or input file you seek, you will in fact be downloading all of the code and input files found in the book. From that point on, you can choose which one you seek and proceed from there. In order to download the data file NMSTATEDATA4.2[1] located in Chapter 5, click HERE.

ACKNOWLEDGEMENTS

All graphs were generated using GraphPad Prism version 4.00 for Macintosh, GraphPad Software, San Diego California USA, www.graphpad.com. Dennis Shasha's work has been partly supported by the U.S. National Science Foundation under grants IIS-0414763, DBI-0445666, N2010 IOB-0519985, N2010 DBI-0519984, DBI-0421604, and MCB-0209754. This support is greatly appreciated. The authors would like to thank Radha Iyengar, Rowan Lindley, Jonathan Jay Monten, and Arthur Goldberg for their helpful reading. In addition, copyeditor Sara Kreisman and compositor Tim Donar worked under tight time pressure to get the book finally ready.

Introduction

Few people remember statistics with much love. To some, probability was fun because it felt combinatorial and logical (with potentially profitable applications to gambling), but statistics was a bunch of complicated formulas with counter-intuitive assumptions. As a result, if a practicing natural or social scientist must conduct an experiment, he or she can't derive anything from first principles but instead must pull out some dusty statistics book and apply some formula or use some software, hoping that the distribution assumptions allowing the use of that formula apply. To mimic a familiar phrase: "There are hacks, damn hacks, and there are statistics."

Surprisingly, a strong minority current of modern statistical theory offers the possibility of avoiding both the magic and assumptions of classical statistical theory through randomization techniques known collectively as *resampling*. These techniques take a given sample and either create new samples by randomly selecting values from the given sample with replacement, or by randomly shuffling labels on the data. The questions answered are familiar: How accurate is the measurement likely to be (confidence interval)? And, could it have happened by mistake (significance)?

A mathematical explanation of this approach can be found in the well written but still technically advanced book *Bootstrap Methods and their Application* by A. C. Davison and D. V. Hinkley. We have also found David Howell's web page ⊟ extremely useful. We will not, however, delve into the theoretical justification (which frankly isn't well developed), although we do note that even formula-based statistics is theoretically justified only when strong assumptions are made about underlying distributions. There are, however, some cases when resampling doesn't work. We discuss these later,

Note to the reader: We attempt to present these ideas constructively, sometimes as thought experiments that can be implemented on a computer. If you don't understand a construction, please reread it. If you still don't understand, then please ask us. If we've done something wrong, please tell us. If we agree, we'll change it and give you attribution.

Contents

CHAPTER 1

The Basic Idea

Suppose you want to know whether a coin is fair[1]. You toss it 17 times and it comes up heads all but 2 times. How might you determine whether it is reasonable to believe the coin is fair? (A fair coin should come up heads with probability 1/2 and tails with probability 1/2.) You could ask to compute the percentage of times that you would get this result if the fairness assumption were true. Probability theory would suggest using the binomial distribution. But, you may have forgotten the formula or the derivation. So, you might look it up or at least remember the name so you could get software to do it. The net effect is that you wouldn't understand much, unless you were up on your probability theory.

The alternative is to do an experiment 10,000 times, where the experiment consists of tossing a coin that is known to be fair 17 times and ask what percentage of times you get heads 15 times or more (see Fig. 1.1). When we ran this program, the percentage was consistently well under 5 (that is, under 5%, a result often used to denote "unlikely"), so it's unlikely the coin is in fact fair. Your hand might ache from doing this, but your PC can do this in under a second.

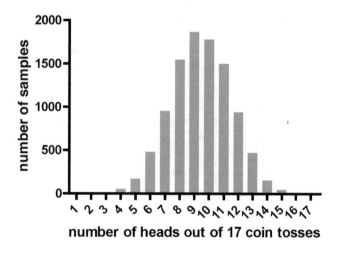

Figure 1.1:   Coin toss.

---

1.   We take this coin example from David Howell.
      See http://www.uvm.edu/~dhowell/StatPages/Resampling/philosophy.html.

Here is an example run of the Coinsig.py code:

```
9 out of 10,000 times we got at least 15 heads in 17 tosses.
Probability that chance alone gave us at least 15 heads in 17 tosses is 0.0009 .
```

Here is a second example.
Imagine we have given some people a placebo and others a drug. The measured improvement (the more positive the better) is:

```
Placebo: 54 51 58 44 55 52 42 47 58 46
Drug: 54 73 53 70 73 68 52 65 65
```

As you can see, the drug seems more effective on the average (the average measured improvement is nearly 63.7 (63 2/3 to be precise) for the drug and 50.7 for the placebo). But, is this difference in the average real? Formula-based statistics would use a *t*-test which entails certain assumptions about normality and variance; however, we are going to look at just the samples themselves and *shuffle* the labels.

The meaning of this can be illustrated in the following table—in which we put all the people— labeling one column 'Value' and the other 'Label' (P stands for placebo, D for drug).

Value	Label
54	P
51	P
58	P
44	P
55	P
52	P
42	P
47	P
58	P
46	P
54	D
73	D
53	D

70	D
73	D
68	D
52	D
65	D
65	D

Shuffling the labels means that we will take the P's and D's and randomly distribute them among the patients. (Technically, we do a uniform random permutation of the label column.)

This might give:

Value	Label
54	P
51	P
58	D
44	P
55	P
52	D
42	D
47	D
58	D
46	D
54	P
73	P
53	P
70	D
73	P
68	P
52	D
65	P
65	D

In Fig. 1.2, we can then look at the difference in the average P value vs. the average D value. We get an average of 59.0 for P and 54.4 for D. We repeat this shuffle-then-measure procedure 10,000 times and ask what fraction of time we get a difference between drug and placebo greater than or equal to the measured difference of 63.7 - 50.7 = 13. The answer in this case is under 0.001. That is less than 0.1%. Therefore, we conclude that the difference between the averages of the samples is real. This is what statisticians call *significant*.

Let's step back for a moment. What is the justification for shuffling the labels? The idea is simply this: if the drug had no real effect, then the placebo would often give more improvement than the drug. By shuffling the labels, we are simulating the situation in which some placebo measurements replace some drug measurements. If the observed average difference of 13 would be matched or even exceeded in many of these shufflings, then the drug might have no effect beyond the placebo. That is, the observed difference could have occurred by chance.

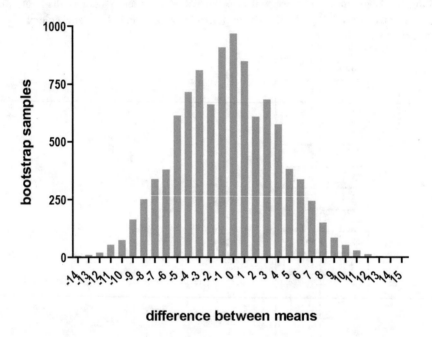

difference between means

Figure 1.2:   Difference between means.

To see that a similar average numerical advantage might lead to a different conclusion, consider a fictitious variant of this example. Here we take a much greater variety of placebo values: 56 348 162 420 440 250 389 476 288 456 and simply add 13 more to get the drug values: 69 361 175 433 453 263 402 489 301 469. So the difference in the averages is 13, as it was in our original example. In tabular form we get the following.

Value	Label
56	P
348	P
162	P
420	P

440	P
250	P
389	P
476	P
288	P
456	P
69	D
361	D
175	D
433	D
453	D
263	D
402	D
489	D
301	D
469	D

This time, when we perform the 10,000 shufflings, in approximately 40% of the shufflings; the difference between the D values and P values is greater than or equal to 13. So, we would conclude that the drug may have no benefit — the difference of 13 could easily have happened by chance.

All code was tested using Python 2.3.

download code and input files

Here is an example run of the Diff2MeanSig.py code, using the first data set from this example as input:

```
Observed difference of two means: 12.97
7 out of 10,000 experiments had a difference of two means greater than or
equal to 12.97 .
The chance of getting a difference of two means greater than or equal to 12.97 is
0.0007.
```

In both the coin and drug case so far, we've discussed statistical significance. Could the observed difference have happened by chance? However, this is not the same as importance, at least not always. For example, if the drug raised the effect on the average by 0.03, we might not find this important, even if the result is statistically significant. That is, the first question you should ask when someone tells you an effect is statistically significant is: "Yes, but how large is the effect?" Perhaps what is being measured here is survival. Say average survival on the placebo is 5 years, and that the drug increases survival on average by 3 days. The difference between 5 years and 5 years and 3 days may be significant, but it is not a large effect.

To get a feeling for this question of importance, we will use the notion of a *confidence interval*. Intuitively, the confidence interval of an imperfectly repeatable measurement is defined by the range of values the measurement is likely to take. In resampling statistics as in traditional statistics, this range is commonly defined as the middle 90% (or sometimes 95%) of the possible values. If you've been following carefully so far, you will guess that the set of possible values will be based on repeated random samples of some sort. In the drug case, we will take many samples from the patient data we have and then look at the difference between the average drug improvement and the average placebo improvement. We'll look at the range of these differences and compute the confidence interval. This technique is called *bootstrapping*.

Here's the method: we create new samples of the same size as the original by choosing values from the original sample "uniformly at random and with replacement."

Let's break down the phrase. "Uniformly at random" means each new sample element is chosen from the original sample in such a way that every original sample element has the same chance of being picked. "With replacement" means that even though an original sample element has been picked, its chance of getting picked again remains the same. Simply put, in forming a new sample (called a bootstrap sample), we choose uniformly at random on the original sample and may choose some elements twice or more and some elements no times at all.

Let's recall our original data regarding drugs and placebos:

```
Placebo: 54 51 58 44 55 52 42 47 58 46
The average is: 50.7.
```

NOTE:
As we will see later, this is in fact not enough data to justify the confidence interval procedure, but is used for easier illustration.

```
Drug: 54 73 53 70 73 68 52 65 65
The average is: 63.7.
```

We subtract the placebo average from the drug average, yielding 63.7 - 50.7 = 13.

Our question now will be: "What is the 90% confidence interval of difference in the averages between the drug patients and placebo?" We answer this with experiments of the form: take a bootstrap sample of the placebo patients and compute the average; take a bootstrap sample of the drug patients and compute the average; then subtract the placebo average from the drug average. When we do this 10,000 times (the rule of thumb for bootstrapping is 1,000 times, but to increase the probability of capturing a wider range of values, we advocate increasing this to 10,000), we get many differences.

Here is a typical experiment in which a bootstrap sample of the placebo values is (note that 54 and 55 are repeated a few times, but 52 never appears):

```
55 54 51 47 55 47 54 46 54 54
The average is: 51.7.
```

Here is a bootstrap of the drug values:

```
68 70 65 70 68 68 54 52 53
The average is: 63.1.
```

We subtract the placebo average from the drug average, yielding 63.1 - 51.7 = 11.4.

When we repeated such an experiment 10,000 times and performed the subtraction each time, the lowest difference was -0.46 (the placebo is a tiny bit more effective than the drug). The highest was 23.4 (the drug is much more effective than the placebo). A more interesting range is the value 5% from the lowest and 95% from the lowest (percentile 5% and 95%). That is, arrange the differences in sorted order from smallest to largest and pick the differences that are at position 500 (500 is 5% of 10,000) and the difference at position 9,500. That is the 90% confidence interval. In our experiments, this yields a range of 7.81 to 18.11. That is, 90% of the time, drugs should yield a value that is 7.81 to 18.11 more than the placebo.

All code was tested using Python 2.3.
<u>**download code and input files**</u>

Here is an example run of the Diff2MeanConf.py code:

```
Observed difference between the means: 12.97
We have 90.0 % confidence that the true difference between the means is between:
7.81 and 18.11
```

<u>Confidence interval</u>s may vary vastly more for social/cultural phenomena than for physical/biological ones. Suppose we have 20 people and we compute their average income. Most have an annual income in the multi-thousand dollar range, but one person has an income of a billion dollars. The average will therefore be something like $50 million, even though most people don't make that much.

What we might be interested in is how far that average value varies if this were in fact a typical sample. Using **bootstrapping** we might have incomes (measured in thousands) as follows:

```
200 69 141 45 154 169 142 198 178 197 1000000 166 188 178 129 87 151 101 187 154
```

This gives an average of 50,142 thousands.
Now if we use the bootstrap, here is another sample:
```
151 154 166 188 154 101 1000000 129 188 142 188 129 142 188 151 87 200 178 129 166
```

This has an average of 50,146 thousands.

Another one:

154 87 178 151 178 87 154 169 187 129 166 154 154 166 198 154 141 188 87 69

This has an average of only 147 thousands (because the billionaire is missing).

Another one:

69 166 169 142 188 198 154 45 187 166 87 154 1000000 87 151 166 101 154 1000000 166

This has an average of 100,127 thousands because the billionaire is present twice.

The net effect is that we are going to get a wide variety of averages. In fact, when we ran the bootstrap 10,000 times on our PC, we obtained a low average of 114 thousands and a high average of 300,118 thousands. But these highs and lows are not so interesting because they vary a lot depending on how many times the billionaire happens to appear. A more interesting range is the value 5% from the lowest and 95% from the lowest (percentile 5% and 95%). That is, arrange the averages in sorted order from smallest to largest and pick the average that is at position 500 (500 is 5% of 10,000) and the average at position 9,500.

On our PC, the 5th percentile is 138 thousands and the 95th percentile is 200,130 thousands. Because 95 - 5 = 90, this is the 90% confidence interval. Because of this vast range, we'd probably conclude that the average is not very informative.

When data has such extreme *outliers*, there are sometimes reasons for ignoring them, such as a faulty meter reading. If there is a good reason to ignore the billionaire in this case, then we get a 90% confidence interval of about 132 to 173 thousands, a much more narrow range of expected values. Unfortunately, outliers are ignored incorrectly sometimes, so one must be careful. Also, as Nassim Taleb points out persuasively, outliers are much more common in human constructs (like income or inflation) than in natural phenomena (like rainflow) 🗗. He gives a particularly convincing example: in the early 1920s, inflation in Germany caused the exchange rate from German Marks to U.S. dollars to go from 3 to a dollar to 4 trillion to a dollar (that is a statistical impossibility under the normal distribution assumption, which illustrates why blind application of the normal distribution can be dangerous).[2]

If you've been following this carefully, you might now wonder "If I have a confidence interval, what more does significance bring to the party?" To answer this intuitively, consider a simple example in which you have just one element of group A having value 50 and one element of group B having value 40. The confidence interval using replacement will say that the difference is always 10. But intuitively this is way too little data. The significance test (in which one permutes the group labels) will show that half the time one will get, just by random chance, a difference as big as the observed one.

If all this seems easy, that's good. Several studies indicate that resampling is easier for students to learn and they get correct answers more often than using classical methods 🗗.

---

2.   Why do we prefer confidence intervals based on the bootstrapping method to traditional confidence intervals based on the standard deviation of the data? First, because we don t want to have to make the assumption that the underlying distribution is normal. Second, because many distributions are in fact skewed. For example, if we want to know the average salary of a population and we know the salaries of a sample, we expect salaries to be positive, whereas the average less the standard deviation might in fact be negative. Bootstrapping looks at the data that is present.

Chapter 1 Exercises

The data sets you need for these exercises will be at http://cs.nyu.edu/cs/faculty/shasha/papers/stateasyExerciseData.zip

1. Run the software on a different set of placebo/drug experiments (Diff2MeanCh1Ex1.vals) and compute a confidence interval and a statistical significance. What are they?

    Hint: In Diff2MeanCh1Ex1.vals: each drug value is exactly 2 more than a placebo value.

2. In Diff2MeanCh1Ex2.vals we have duplicated the placebo/drug data from exercise 1 to see how the confidence interval and statistical significance change. How do they?

3. Consider a year in which there are daily closing prices from stocks A and B. Compute the correlation of closing prices of stocks A and B. Also what is your confidence that the correlation is positive and what is the confidence interval of the correlation?

The file RegressionCh1Ex3.vals contains JP Morgan Chase & Co and Bank of America Corporation closing prices for every day in 2009.

Chapter 1 Solutions

The data sets you need for these exercises will be at http://cs.nyu.edu/cs/faculty/shasha/papers/stateasyExerciseData.zip

2. In Diff2MeanCh1Ex2.vals we have duplicated the placebo/drug data from exercise 1 to see how the confidence interval and statistical significance change. How do they?

    Solution:
    Significance:
    Observed difference of two means: 2.00
    68 out of 10000 experiments had a difference of two means greater than or equal to 2.00. The chance of getting a difference of two means greater than or equal to 2.00 is 0.0068.

    Significance improves when we double the sample size. We can see this because the chance of getting a difference of 2 or greater under random permutation decreases.

    Confidence:
    Observed difference between the means: 2.00
    We have 90.0 % confidence that the true difference between the means is between: 0.74 and 3.28

    Our confidence interval got a little tighter. The reason is that with a larger sample size outliers have less of an impact (even though they still occur in the data with the same frequency). Remember here we bootstrap the samples to generate new samples so not every value is included in every bootstrapped sample.

CHAPTER 2

Pragmatic Considerations When Using Resampling

This material is based on Sec. 2.6 of the Davison and Hinkley book *Bootstrap Methods and their Application*, and Philip Good in his book *Resampling Methods*. It concludes with a short primer on statistical jargon to which you may want to refer in your reading.

1. **Bootstrapping** may underestimate the size of the **confidence interval** when the sample is small (as a rule of thumb, according to Philip Good, under 100). In that case, a **significance** test may work better. Significance/**shuffle**/permutation tests (when one shuffles the label) can be used with as few as three data points of each type. This implies that there may be cases when one wants to reformulate a question as a significance test when there are few data items. For example, to evaluate whether a treatment T1 encourages, say, growth more than treatment T2, one could use three strategies:

    i. Find the 95% **bootstrap confidence interval** of the difference in the means between T1 and T2.

    ii. Label each data point with the treatment it came from. Then do a **significance** test by shuffling the labels and seeing how often the difference in the means exceeds the observed difference.

    iii. **Shuffle** the labels as in (ii) but choose the data points by sampling with replacement. (A hybrid significance approach.)

    Option (i) implicitly assumes there is a difference and simply attempts to find the confidence interval of its magnitude. As such, it does what statisticians call "starting with the **alternative hypothesis**." Options (ii) and (iii) start with the **null hypothesis** (the treatment doesn't matter). The **p-value** calculated in their case is the probability that the treatment doesn't matter but that the difference one sees is as great as that observed. That's why they constitute a significance approach.

    To see why (i) differs in its implicit assumption from the other two, suppose that we had the degenerate case of just one data value from each treatment, e.g., the T1 value is 55 and the T2 value is 48. Option (i) would always find the same difference between the means, in this case 7, so the 95% confidence interval would range from 7 to 7. Options (ii) and (iii) would reshuffle the labels so about half the time the labels would be reversed, so the **p-value** of a difference this great would be about 0.5, obviously not significant.

    Pragmatically, therefore, you should start by deciding whether the null hypothesis should be rejected before measuring the magnitude of the difference. If the variances of the two samples are different, then using (iii) is probably best.

2. **Bootstrapping** should not be used to find the maximum value of an underlying population (e.g., to try to find the tallest person in Holland, it would not do to sample 1,000 Dutch people). This holds also if one wants to find the kth largest value. On the other hand, bootstrapping can be used to estimate the cutoff demarking the 10% largest values (also known as the highest decile) and to get **confidence intervals** for those values.

3. Bootstrapping also should be used with great caution whenever there are outliers in a sample, e.g., values very different from the others that can radically change the evaluation of the statistic. If outliers are a problem, then a *rank transformation* might work better (e.g., convert every salary to a salary rank as in 1 for the highest, 2 for the second highest, etc.).

    To locate the outliers (see http://en.wikipedia.org/wiki/Outlier), start by computing the difference between the third quartile value q_3 (the value below which 75% of the data lies) and the first quartile value q_1 (the value below which 25% of the data lies). Call this the inter-quartile range IQR. An outlier is any value less than q_1 - 3IQR or greater than q_3 + 3IQR. A temptation is to discard outliers. As mentioned earlier, throwing out outliers blindly can lead one to miss important phenomena (holes in the ozone layer, people with extraordinarily high income, stock market crashes, and so on).

    More sophisticated statistical approaches for handling outliers use a technique known as "Robust Statistics" described nicely in the texts by Huber or Hampel et al. The consensus among the professionals, however, is that outliers should be thought about carefully or transformed into ranks rather than statistically eliminated.

4. Neither bootstrapping nor significance should be used when the sample is not a representative one. For example, if you are doing a survey and the yield (the fraction of people called who respond) is very low, then the sample may be biased. Whether it is or not depends on whether the reason for refusal is relevant to the quantity you're measuring. For example, if people opt out of a medical study because they find out they are getting a placebo, then any conclusions about outcomes may be biased in favor of the drug. It is the possible lack of representativeness of exit polling that may explain why voting results at the ballot box seem to differ from exit polling estimates. Lack of representativeness is a problem for any statistical technique.

    Sometimes there can be experimental bias. Philip Good observes that when placing subjects into treatment pools, researchers tend to put the most promising ones into the first pool (e.g., the healthiest plants into the first garden). It is essential, therefore, to randomly distribute values among the various treatment pools. The pseudocode if you have *n* subjects and *k* treatments is:

    ```
    for subject 1 to n
      choose a number x and at random between 1 and k
      put the subject in treatment x unless x is full
      in which case you choose again until you find a treatment
          that is not full
    end for
    ```

5. Resampling should be used with care when the data exhibits serial dependence. For example, permuting individual elements in a musical melody or stock price time series may give erroneous results for the simple reason that the value at each time point depends to a great extent on the value at the previous time point. In such a case, the basic assumption of resampling—that values in the sample don't depend on the values around them—doesn't hold. A practical fix for this, if the sample is large enough, is to take non-overlapping long consecutive subsequences of data and then do resampling on entire subsequences. This technique preserves the local dependencies of the data except at the borders of the subsequences. Tomislav Pavlicic of DE Shaw offered us the following case history: "I successfully used resampling for mutually

dependent data. It did require creating 50,000 virtual subjects each with 100 mutually dependent data points and then performing resampling on non-overlapping (longish) subsamples. It
worked! (The alternative would have required the kind of expertise in combinatorics which
was way above my head.)"

Victor Zhu pointed out another (advanced) technique for evaluating significance when there is
serial dependence: "One technique I have seen in permutation for time series is to calculate
the wavelet coefficients of each time series, to permute the labels associated with each coefficient [so the first wavelet coefficient from time series x may become the first wavelet coefficient from time series y] and then to reconstruct the several time series. This is a way to test
the significance of correlations.

"Another question I have often encountered in interviews is how to tell if one trading strategy
is better than another, or is it just due to luck. Try a shuffle test."

6. *Rank transformations* can be very useful sometimes. For example, suppose you are calculat-
   ing the economic value of going to college. You look at the salary of all people who graduated
   from college as compared to all people who didn't. The trouble is that a few outliers such as
   Bill Gates might significantly influence the result. So, you might decide to order all salaries
   and replace each person's salary by his or her rank (say that Gates would be 1). Then the question is: what is the average rank of college graduates vs. non-college graduates? If lower for
   college graduates, then college has some economic benefit in the sense that college graduates
   tend to be among the higher earners. (Remember that rank 1 is the richest individual.) In
   detail, this analysis would involve associating with each individual a rank and a label (college
   or non-college). Compute the average rank of the individuals having each label. Then perform
   a significance test that involves randomly *shuffling* the labels and measuring the difference in
   the means. By doing this repeatedly, you will determine how often the average college gradu-
   ate rank is less than the average non-college rank. This would give a p-value (or significance
   value). Note that this transformation addresses the question: "Do college graduates tend to be
   among the higher earners?" rather than "Do college graduates have a higher mean salary?"

   Transformations can also involve simple algebra. For example, if you want to test the propo-
   sition that Dutch men are more than five centimeters taller than Italian men, you can make
   the null hypothesis be that Dutch men are five centimeters taller. To do this, first subtract five
   centimeters from the heights of the Dutch men in your sample. Now you have men repre-
   sented by a pair (height, nationality), where the heights have been adjusted for the Dutch
   men. The null hypothesis is now that the average height difference should not differ from 0.
   Suppose that the average (adjusted) height among the Dutch men is in fact greater than the
   average height among the Italian men. Suppose that after the transformation Dutch men were
   on average still 4 centimeters taller (so 9 centimeters on the original). Suppose further that
   under the null hypothesis after the transformation, shuffling the nationality labels seldom
   yields this outcome. If that is true, then the observed extra 4 centimeters has a low p-value so
   the null hypothesis is unlikely to hold. That is, the height difference is likely to be greater
   than 5 centimeters.

7. For the most part, we concentrate on the use of resampling to estimate the significance of a
   result (using shuffling) or the size of a result (using bootstrapping) regardless of the underlying distribution. An important alternative use of resampling is to estimate the probability
   that a sample comes from a particular distribution. We did this *earlier* when we asked how
   likely a set of 17 flips of a fair coin would be to yield 15 heads or more. The basic concept is

to use bootstrapping from the assumed distribution and then to see how often the observed outcome materializes.

8. The *power of a test* is the probability that the test will reject the null hypothesis if the null hypothesis is false. The power increases the larger the size of the sample. Given a desired power and either a distribution D reflecting your prior belief or an *empirical distribution* (a distribution based on an experiment), you can determine the size of the sample you need to achieve that power. The idea is simple: assume that D is true and say you are testing some statistic such as the mean M and you want to know how big a sample you'll need to conclude 90% of the time (i.e., with a power of 90%) that $M > 50$ with a p-value < 0.05.

Specify a sample size S. We will see whether S is large enough to give us the power we want to achieve the alternate (i.e., non-null) hypothesis, that $M > 50$.

```
Try the following 1,000 times:
   draw 10,000 samples of size S from the empirical distribution
                with replacement
            see whether M is over 50 at least 95% of the time
            If so, count this as a reject of the null hypothesis
```

The power is the fraction rejects/1,000. If the power is too low (less than 0.9 in this example), you'll need a bigger sample.

For example, suppose your empirical distribution indicates that the data has a probability of 0.4 of having the value 40 and a probability of 0.6 of having the value 70. You want a power of 0.8 that you will find a mean greater than 50 with a *p*-value under 0.05.

```
Try 1,000 times with a sample size S:
   take 10,000 samples of size S from this distribution
   if 95% or more of the time, one gets a mean over 50,
         then the number of rejects increases by 1
```

The sample size is big enough if the number of rejects divided by $1,000 > 0.8$. Otherwise try a larger S and repeat. See also 🗗.

Note that we have selected a single sample mean as our test statistic, but this method can be used for multiple data sets and other statistics. The point is to use this same strategy: assume a distribution then try a sample size by repeatedly drawing a sample from that distribution of that size.

For example, suppose you believe that treatment T1 obeys some distribution D1 and T2 obeys D2 and the mean of D1 is greater than the mean of D2. You want to know whether taking a sample of S points from each will often cause you to reject the null hypothesis that the mean of T1 equals the mean of T2 with a *p*-value of 0.05.

```
Do the following 1,000 times
   Do 10,000 times: take a sample of size S from presumed
      distribution D1 and another sample of size S from presumed distribution D2
      and see whether mean of the sample from D1 >  mean of the sample from D2
   If this is true 95% of the time, then count this as a reject of the null
hypothesis.
```

The power is the number of rejects divided by 1,000.

In order to find the correct sample size more efficiently, we may perform the outer loop far fewer times. If after 100 iterations of the outer loop the power looks way too low, then we increase the sample size immediately.

9.   Multi-factor designs and *blocking*[1]. You want to test different dosages of a fertilizer on a field. You are worried, however, that the properties of the field might swamp the effect of the fertilizer. On the other hand, it is not practical to give two plants that sit right next to one another a different fertilizer, so large scale randomization is not possible. For this reason, you might use a *Latin square design* so similar dosages (Lo, Med, Hi) of fertilizer are not applied to neighboring fields.

	east	—	west
north	Hi	Med	Lo
—	Lo	Hi	Med
south	Med	Lo	Hi

This is better than a randomized design because a randomized design on a three-by-three square might give a very biased outcome like:

	east	—	west
north	Hi	Hi	Lo
—	Lo	Hi	Med
south	Med	Lo	Med

When doing a <u>shuffle</u> test in this setting, it is helpful to shuffle between one Latin square and another. That is, see how extreme the outcome is in the original setting compared to how it would be if we shifted the dose labels, e.g., one shuffle might shift the labels to:

	east	—	west
north	Lo	Hi	Med
—	Med	Lo	Hi
south	Hi	Med	Lo

1.   This example is from Philip Good.

Chapter 2 exercises

The data sets you need for these exercises will be at http://cs.nyu.edu/cs/faculty/shasha/papers/ stateasyExerciseData.zip

1. Consider a set of salaries and whether people had gone to college or not. Using just that data, evaluate whether going to college gives a higher salary with p-value < 0.05.

    Diff2MeanCh2Ex1Unranked.vals: High school graduate salaries verses college graduate salaries (no graduate school).

2. After transforming that data to ranks, evaluate the same question.

3. Consider again exercise 1 from chapter 1. We have 27 drug samples and 27 placebo samples. Is this a big enough sample to establish with a 95% confidence that the drug is significantly better than the placebo? If not, how much more data does one need to do so? We want a 95% probability that we will reject the null hypothesis if the null hypothesis should be rejected. (Hint: see the power of a test section.)

    Hint: In other words, if the current sample size is S and the average drug result is better than the average placebo result, how many more replicas of S would be needed for us to believe that the drug is better than the placebo throughout the 95% confidence interval? Suppose that, in the course of using the bootstrap, we sorted the difference drug average - placebo average. If every number from the bottom 2.5% of those differences on up always gave a positive value, then we have achieved what we want.

4. You have four different treatments: none, A alone, B alone, A and B together

    You have 16 fields: four are dry-hilly, four are wet-hilly, four are dry-flat, four are wet-flat. Give two Latin square designs that work for this. Call the four dry-hilly fields dry-hilly1, dry-hilly2, dry-hilly3, and dry-hilly4. And similarly for the four wet-hilly fields and so on.

Chapter 2 Solutions

The data sets you need for these exercises will be at http://cs.nyu.edu/cs/faculty/shasha/papers/ stateasyExerciseData.zip

2. After transforming that data to ranks, evaluate the same question.

    Solution:
    Rank 1 will be the highest salary, rank 2 the second highest etc. If multiple salaries are identical, they are all assigned the same rank. If there are x people with salaries u and r is the next unused rank, all x people are assigned rank $(r + ((r - 1) + x)) / 2$. The next unused rank is then $r + x$. For example:

salary   rank
12499   4
12500   5.5
12500   5.5
15000   6

Two people have salary 12500. So $x = 2$. The next unused rank is 5, so $r = 5$. Both people with salary 12500 will be assigned rank $(5 + ((5 - 1) + 2)) / 2 = 11 / 2 = 5.5$. The next unused rank is 7.

Now when we run Diff2MeanSig.py on Diff2MeanCh2Ex1Ranked.vals, we get:

Observed difference of two means: 10.21
62 out of 10000 experiments had a difference of two means greater than or equal to 10.21 .
The chance of getting a difference of two means greater than or equal to 10.21 is 0.0062 .

Our p-value is 0.0062, which is still less than 0.05. When we ran Diff2MeanConf.py on the above input file we got:

Observed difference between the means: 10.21
We have 90.0 % confidence that the true difference between the means is between: 3.63 and 16.57

When we run Diff2MeanConfCorr.py on the ranked input file we get:

Observed difference between the means: 10.21
We have 90.0 % confidence that the true difference between the means is between: 3.46 and 16.60

The difference of course is now measured in salary ranks and not dollars.

4. You have four different treatments: none, A alone, B alone, A and B together

You have 16 fields: four are dry-hilly, four are wet-hilly, four are dry-flat, four are wet-flat. Give two Latin square designs that work for this. Call the four dry-hilly fields dry-hilly1, dry-hilly2, dry-hilly3, and dry-hilly4. And similarly for the four wet-hilly fields and so on.

Solution:
Here is a first Latin square design.

dry-hilly1 none
dry-hilly2 A
dry-hilly3 B
dry-hilly4 AB
wet-hilly1 none

wet-hilly2 A
wet-hilly3 B
wet-hilly4 AB
dry-flat1 none
dry-flat2 A
dry-flat3 B
dry-flat4 AB
wet-flat1 none
wet-flat2 A
wet-flat3 B
wet-flat4 AB

Here is a second one (just rotation for each field type):

dry-hilly1 AB
dry-hilly2 none
dry-hilly3 A
dry-hilly4 B
wet-hilly1 AB
wet-hilly2 none
wet-hilly3 A
wet-hilly4 B
dry-flat1 AB
dry-flat2 none
dry-flat3 A
dry-flat4 B
wet-flat1 AB
wet-flat2 none
wet-flat3 A
wet-flat4 B

CHAPTER 3

Terminology

Statistics has a lot of terminology. For the most part, we try to avoid it. On the other hand, you will hear other people use it, so you should know what it means.

acceptance region

The lower limit of the *p*-value at which point one accepts the <u>null hypothesis</u>. For example, if the cutoff is 5%, then one accepts the null hypothesis provided the *p*-value is 5% or greater. As Philip Good points out, academics commonly present the *p*-value without further interpretation. Industrial people have to decide. For example, suppose you are testing a drug and your null hypothesis is that it is non-toxic. You conduct a trial with 200 people, 100 of whom get the drug. If 23 people fall sick using the drug and only 21 when not, you'd probably reject the drug even though the *p*-value is quite high (and therefore there is no statistical reason to reject the null hypothesis that the drug is non-toxic). See <u>rejection region</u> and *<u>p-value</u>*.

alternate hypothesis

See <u>null hypothesis</u>.

blocking effect

If your independent variable is X and there is another possible confounding factor F and you test $X = x_1$ and $F = f_1$ and a second test $X = x_2$ and $F = f_2$, then the confounding factor F may swamp the effect of the independent variable.

block design

This is to avoid the <u>blocking effect</u> as follows: if F might be a confounding factor to the independent variable X, then make sure that each setting of X is tested with every value of F. Example: you want to test the wearability of a different shoe type so you have shoe types s_1 and s_2. Confounding factor is the person. So, if you have people $p_1, p_2, ..., p_n$, it would be bad to give s_1 to the first person and s_2 to the second one, and so on. Instead, give s_1 and s_2 to p_i with a random selection of left and right foot (another possible confounding factor). Thus for each s_i, you want to use all possible values of the confounding factor approximately equally. The <u>Latin square approach</u> quoted earlier is another technique to avoid a blocking effect.

distribution-free test

Is one whose result does not depend on any assumption about the underlying distribution. The <u>significance</u> test based on <u>shuffling</u> is distribution-free. The only assumption it makes is that if the <u>null hypothesis</u> did hold then one could rearrange the labels without affecting the underlying distribution of the outcomes. For such tests, the null hypothesis is usually exactly this (informally, the labels, e.g., treatments, don't matter). So this is not much of an assumption. The <u>bootstrap</u> is also distribution-free.

exact test

Is one that correctly assigns a *p*-value without regard to the distribution. See <u>distribution-free</u> and <u>*p*-value</u>.

non-parametric

See <u>distribution-free</u>.

null hypothesis

We have a null (boring) hypothesis (H_0) which is that whatever we see is due to chance alone, and an alternate (exciting) hypothesis that whatever we see is not due to chance alone, but something else. One hypothesis must be true and they cannot both be true. The question you have to decide, based on your data, is whether your null hypothesis is supported by your data, or if the null hypothesis should be rejected in favor of your exciting hypothesis.

one-tailed

See <u>tails</u>.

_p_-value

In a <u>significance</u> test the *p*-value is the probability of observing a summary from the data at least as extreme as the one calculated from the data under the assumption that the <u>null hypothesis</u> is true. For example, if we flip a supposedly fair coin 100 times, the *p*-value would be greater for the outcome 59 heads and 41 tails assuming the coin is fair than for the outcome 89 heads and 11 tails under the same assumption. This could be calculated by doing 10,000 bootstraps assuming a fair coin and counting the number of times the result was more extreme than this outcome. Call that number N. The *p*-value would then be $N/10,000$. (Of course, the binomial theorem could also be used.) Similarly, if the null hypothesis is that a treatment doesn't matter and a sample S shows that the treatment results differ by 30% from the placebo results, then the *p*-value of this null hypothesis can be computed by <u>shuffling</u> the labels (treatment/non-treatment) on S 10,000 times and counting how often the result is at least as extreme as the result observed. Call that number N. The *p*-value would then be $N/10,000$.

parametric method

Is one whose <u>significance</u> estimates depend on a specific distribution.

parametric bootstrap

Is <u>bootstrapping</u> according to a known distribution (e.g., the <u>fair coin assumption</u>) to see whether it is likely that some observed outcome is likely to have resulted from that distribution.

permutation test

Is what we are calling a <u>significance</u> or <u>shuffle</u> test.

power of a test

Is the probability that the test will reject the <u>null hypothesis</u> if the null hypothesis should be rejected. One can get a power of 1, if one always rejects the null hypothesis, but this fails to be *conservative* (i.e., one rejects too often). In general, the greater the <u>*p*-value</u> that separates the <u>rejection</u> from the <u>acceptance region</u>, the greater the power. The power also increases the larger the size of the sample. Given a desired power and either a distribution D reflecting your prior belief or an __empirical distribution__, you can determine the size of the sample you need to achieve that power as explained in the pragmatic consideration section (Chapter 3) above.

rank transformation

Is a __transformation__ on a collection of numbers in which the numbers are sorted and then each number is replaced by its order in the sorted sequence. The sorting can be ascending or descending. You have to specify which.

rejection region

__p-value__ below which you reject the __null hypothesis__. For example, the p-value that in 100 flips of a fair coin, there are 89 heads is so low that we would reject the hypothesis of fairness. Suppose we were conducting an experiment of the effectiveness of a drug and we conducted a trial of 200 people, 100 of whom received the drug and 100 not. The null hypothesis is that the drug is ineffective. If 23 of the 100 who received the drug got better but only 21 of the 100 who didn't receive the drug got better, then the p-value would be high, so there would be no reason to try the drug. See __acceptance region__ and __p-value__.

tails

Suppose you are testing two treatments — yours and the competition. In a sample, yours seems to perform better. You want to see if the improvement is __significant__. The __null hypothesis__ is that the other treatment is as good as yours or better. The __alternative__ is that yours is better (a one-tailed alternative). So, you do a significance test. If it is very rare for the __shuffled__ labels to show an advantage for your label as great or greater than the one you see, then you reject the null. Suppose now that you are a consumer watch organization and you are skeptical that there is any difference at all between the two treatments. You take a sample and you observe a difference with treatment A giving a higher value than treatment B. Your null hypothesis is that there is no difference. You reject the null if the probability that a shuffle gives as large a magnitude difference *in either direction* is small. This is called a two-tailed test.

transformation

Is any change to the data (e.g., taking the log of all values) that is called for either by the application or to use a certain test. __Earlier__ in this lesson, we subtracted 5 centimeters from the height of each Dutch male in a sample to be able to test the null hypothesis that Dutch men were on average 5 centimeters taller than Italian men. Log transforms are useful when the difference between two treatments is expected to be multiplicative (e.g., a factor of 2) rather than additive.

two-tailed

See __tails__.

type I error

A type I error is thinking that the null hypothesis is false when it's in fact true (*false positive*). The p-value represents the probability of a type I error when you assert the null hypothesis is rejected.

type II error

A type II error is thinking that the null hypothesis is true when in fact it is not (*false negative*). Generally, if we reduce the probability of a type I error, we increase the probability of a type II error.

Chapter 3 Exercises

These exercises are conceptual rather than quantitative. So, no calculation is required.

1. When might you do a rank transformation?
2. If you do a test and the *p*-value of the alternate hypothesis is say less than 0.10, what does that mean exactly?
3. Why do you use shuffling? How is it used? Relate it to a *p*-value. Use a drug vs. placebo scenario as an example.
4. Why do you use bootstrapping? How is it used? Relate it to a confidence interval. Use a drug vs. placebo scenario as an example.
5. When would you use a one-tailed test vs. a two-tailed test?
6. In Nicholas D. Kristof's June 16th, 2010 New York Times op-ed column called "Dad Will Really Like This" he describes tuberculosis sniffing giant rats working in Tanzania developed by a company called Apopo. He says "A technician with a microscope in Tanzania can screen about 40 samples a day, while one giant rat can screen the same amount in seven minutes." What if when the rats were initially tested these were the results:

	Rat signals sample has TB	Rat signals sample is TB free	Total
TB Sample	516	84	600
TB free sample	40	360	400
Total	556	444	1000

Note that there are two types of mistakes here, when the rat indicates the sample has TB and it doesn't (a false-positive) and when the rat signals that the sample is TB is sample free when it isn't (a false-negative). What are the false positive and false negative rates?

7. If you want to prove that your drug has a beneficial effect compared to a placebo, would you want a statistical test that gives more power or less?
8. Without changing the test, how might you increase the power of protocol that uses that test?
9. Relate the terms: false positive, false negative, type 1 error, type 2 error.

Chapter 3 Answers

These exercises are conceptual rather than quantitative. So, no calculation is required.

1. When might you do a rank transformation?

   Answer: When there are some outliers and their values are less important than their rank

2. If you do a test and the p-value of the alternate hypothesis is say less than 0.10, what does that mean exactly?

   Answer: It means that if the null hypothesis is true, then 10% of the time by chance, you would see a test value as extreme as the value you see or more so.

3. Why do you use shuffling? How is it used? Relate it to a p-value. Use a drug vs. placebo scenario as an example.

   Answer: Shuffling is used to test statistical significance. Suppose the drug group has a different effect than the placebo group on some statistic where it has some difference d. In a shuffle test, the assignment of the drug/placebo label to people is shuffled (permuted). Then the statistic is recalculated. The p-value is the number of shuffles where the statistic has a value of d or greater divided by the total number of shuffle experiments. If the p-value is high, then the apparent effect of the drug may be due to chance.

4. Why do you use bootstrapping? How is it used? Relate it to a confidence interval. Use a drug vs. placebo scenario as an example.

   Answer: Suppose again the drug group has a different effect than the placebo group on some statistic where it has some difference d. How much confidence should we have in the specific d value? Bootstrapping attempts to answer this without assuming anything about the sample distribution. In bootstrapping, we repeatedly take random samples of the same size as our original sample, from the original sample. We use these samples to calculate a range of values our test statistic is likely to take. This gives a confidence interval.

5. When would you use a one-tailed test vs. a two-tailed test?

   Answer: You measure some sample statistic (e.g. difference in the mean). Suppose you care only about results that are as extreme or more extreme than your sample statistic in one directon. Then you are concerned about a single tail result.

   If you care about extremes at either end of the range of possibilities, you should use a two-tailed test.

For example, say there is a new diet trend and you want to know if the people who are on it weigh significantly more or less than people who aren't. You look at both tails of the distribution because the diet could have the opposite affect of what people want: it causes weight gain.

6. In Nicholas D. Kristof's June 16th, 2010 New York Times op-ed column called "Dad Will Really Like This" he describes tuberculosis sniffing giant rats working in Tanzania developed by a company called Apopo. He says "A technician with a microscope in Tanzania can screen about 40 samples a day, while one giant rat can screen the same amount in seven minutes." What if when the rats were initially tested these were the results:

	Rat signals sample has TB	Rat signals sample is TB free	Total
TB Sample	516	84	600
TB free sample	40	360	400
Total	556	444	1000

Note that there are two types of mistakes here, when the rat indicates the sample has TB and it doesn't (a false-positive) and when the rat signals that the sample is TB is sample free when it isn't (a false-negative). What are the false positive and false negative rates?

Answer: So in this case, the rats correctly identified 86% of the TB samples. So, 14% of the time when a patient did have TB the rats did not identify it (the false-negative rate). 10% of the TB-free samples they said had TB (the false-positive rate). These rates are about what they found in their preliminary tests (http://arstechnica.com/old/content/2007/09/of-mice-and-mines-trained-rats-search-for-explosives-tuberculosis.ars).

Now what if they had found the following:

	Rat signals sample has TB	Rat signals sample is TB free	Total
TB Sample	84	516	600
TB free sample	360	40	400
Total	444	556	1000

Here the rats signal that the TB sample is TB free 86% of the time and 90% of the time a sample is TB free they say it has TB. Quite likely they have been trained to give a signal that a sample has TB when in fact they don't smell TB at all. In this case it is just a matter of correcting either our interpretation of the signal they give us, or retraining them to give the signal when they do smell TB.

7. If you want to prove that your drug has a beneficial effect compared to a placebo, would you want a statistical test that gives more power or less?

Answer: More. The null hypothesis is that the drug has no effect. Power is the probability that a test rejects the null hypothesis when the null hypothesis should be rejected. The null hypothesis is that the drug has no more effect than the placebo. So, you'd like to increase the power.

8. Without changing the test, how might you increase the power of protocol that uses that test?

   Answer: increase the sample size.

9. Relate the terms: false positive, false negative, type 1 error, type 2 error.

   Answer:

   false positive/type I error - rejecting the null hypothesis when it is in fact true

   false negative/type II error - not rejecting the null hypothesis when in fact it is false

   Four things can happen:

	you reject the null hypothesis	you accept the null hypothesis
null hypothesis is actually true:	false positive/type I error	you are correct
null hypothesis is actually false:	you are correct	false negative/type II error

   Trying to decrease the number of false positives usually means increasing the false negatives and visa-versa. What you choose to do depends on your problem and how costly each error is. For example, in the tuberculosis sniffing rat example from question 6, we might prefer to increase the number of false positives in order to reduce the number of false negatives. That is, we would rather have the rat say someone has TB when they don't (which we will probably figure out with further testing) than have the rat miss someone who actually does have TB.

CHAPTER 4

The Essential Stats

Our approach from now on is to name a statistic, discuss its domain of application, show how to calculate it, give a small example, give pseudo-code for <u>confidence intervals</u> and <u>significance</u> as appropriate, and then a link to code. These can be read in any order.

Mean

A representative value for a group. This is what people are usually referring to when they use the term average.

Difference between Two Means

Used to compare two groups.

Chi-Squared

Usually used to measure the deviation of observed data from expectation or to test the independence of two variables.

Fisher's Exact Test

Used instead of Chi-Squared when one or more of the expected counts for the four possible categories ($2 \cdot 2 = 4$) are below ten.

One-way ANOVA

Used to measure how different two or more groups are when a single independent variable is changed.

Multi-way ANOVA

Used to test the influence of two or more factors (independent variables), on our outcome, when we have two or more groups.

Linear Regression

Used to find the line that best fits data so that it can be used to predict y given a new x.

Linear Correlation

Used to determine how well one variable can predict another (if a linear relationship exists between the two).

Multiple Regression

Used to determine how well a set of variables can predict another.

Multiple Testing

If a statistical test is run many times we expect to see a few values that look significant just due to chance. We discuss several ways to address this problem here.

4.1   Mean

4.1.1   Why and when

If you had just one way to characterize a collection, the mean would probably be the one to use. The mean also has a physical interpretation. If each number in a group is assigned unit weight and rests at its number position on a number line, then the center of gravity is the mean.

4.1.2   Calculate

Sum of all elements in a group/number of elements in the group.

4.1.3   Example

Say we have developed a new allergy medication and we want to determine the mean time to symptom relief. Here we record the time, in minutes, it took 10 patients to experience symptom relief after having taken the medication (if this were a real study, we would have far more patients): 60.2 63.1 58.4 58.9 61.2 67.0 61.0 59.7 58.2 59.8.

   We want a range of values within which we are 90% confident that the true mean lies.

4.1.4   Pseudocode & code

All code was tested using Python 2.3.

<u>download code and input files</u>

Here is an example run of the MeanConf.py code:

```
Observed mean: 60.75
We have 90.0 % confidence that the true mean is between: 59.57 and 62.15
```

4.2   Difference between Two Means

4.2.1   Why and when

Very often someone may claim to improve something and give you before and after data, call them B and A. To evaluate the claim, you might calculate the mean of B and the mean of A and see if the latter is greater than the former (assuming higher values are better). But then two questions arise: (i) Could this improvement have arisen by chance? (ii) How much of a gain should we expect, expressed as a range?

4.2.2   Calculate

Compute the mean of A and the mean of B and find the difference.

4.2.3   Example

See <u>Chapter 1</u>.

4.2.4   Pseudocode & code

All code was tested using Python 2.3.

<u>download code and input files</u>

Here is an example run of the Diff2MeanSig.py code:

```
Observed difference of two means: 12.97

7 out of 10000 experiments had a difference of two means greater than or equal to
12.97 .

The chance of getting a difference of two means greater than or equal to 12.97 is
0.0007 .
```

<u>download code and input files</u>

Here is an example run of the Diff2MeanConf.py and Diff2MeanConfCorr.py code:

```
Observed difference between the means: 12.97

We have 90.0 % confidence that the true difference between the means is between:
7.81 and 18.11
```

4.3   Chi-squared

4.3.1   Why and when

The chi-squared statistic is usually used to measure the deviation of observed data from expectation or to test the independence of two group categories.

First we will use it to test if data has an expected distribution.

4.3.2   Calculate with example

We want to test to see if a die is fair. There are 6 possible outcomes (categories) when the die is rolled: 1, 2, 3, 4, 5, 6. Roll the die 60 times and keep track of the results in a frequency chart. We will also record the expected outcome for each category. We expect each category to have a probability of occurring of 1/6, so if we roll the die 60 times we expect each side to appear 10 times.

Used for *categorized data*.

Do n trials and make a frequency table that contains expected number of hits in each category i as well as the actual number of hits in each category i.

$$\chi^2 = \left(\frac{\left(X_1 - \mathrm{E}(X_1)\right)^2}{\mathrm{E}(X_1)} \right) + \ldots + \left(\frac{\left(X_n - \mathrm{E}(X_n)\right)^2}{\mathrm{E}(X_n)} \right)$$

Category	X_i	$E(X_i)$	$X_i - E(X_i)$	$(X_i - E(X_i))^2$	$(X_i - E(X_i))^2 / E(X_i)$
Description	Observed number of times die landed on side i	Expected number of times die will land on side i	Difference of observed from expected	Difference of observed from expected squared	
1	14	10	4	16	1.6
2	16	10	6	36	3.6
3	6	10	-4	16	1.6
4	9	10	-1	1	.1
5	5	10	-5	25	2.5
6	10	10	0	0	0
Totals	60	60*	0**	-	$\chi^2 = 9.4$

* This total must always be the same as the observed total since it is the number of trials.
** This must always be zero, since if we observed more than we expected in one category, we must have observed less than we expected in others.

Rule of thumb: The expected number of outcomes in each category must be 10 or larger, and we must have at least 2 degrees of freedom.

$$df = n - 1$$

Where n is equal to the number of categories, in this case 6.

There are $n - 1$ degrees of freedom because once we know $X_1, X_2, ..., X_{n-1}$, we know X_n. In our example we have 5 degrees of freedom, and each expected category count is at least 10, so we meet the requirements for a chi-squared test.

If the difference between observed counts and expected counts is large, we reject the hypothesis that the difference between the observed data and expected data is from chance alone, but how do we know when χ^2 is large enough to do this?

We want to determine what the chances are, if the die really is fair, that we get a χ^2 of 9.4 as we did above. If the die is fair would this happen rarely? Would it happen frequently? We figure this out by simulating 60 rolls of a fair die, 10,000 times. Draw 10,000 samples of size 60 and compute chi-squared statistic for each. See how many times the chi-squared statistic is greater than or equal to 9.4. This is our p-value, the chance that, given a fair die, we would get a chi-squared value at least as great as ours.

4.3.3   Pseudocode & code

All code was tested using Python 2.3.

download code and input files

When we ran the ChiSquaredOne.py code on the example above we got:

```
Observed chi-squared: 9.40
963 out of 10000 experiments had a chi-squared difference greater than or equal
to 9.40
Probability that chance alone gave us a chi-squared greater than or equal to 9.40
is 0.0963
```

4.3.4   Calculate with example for multiple variables

Here is another example.

Say we're looking at three income levels and sickness vs. health. Then we might have a table like this:

	Poor	Middle	Rich	Total
Sick	20	18	8	46
Expected	$(46 / 110) \cdot 44 =$ 18.48	$(46 / 110) \cdot 42 =$ 17.64	$(46 / 110) \cdot 24 =$ 10.08	
Healthy	24	24	16	64
Expected	$(64 / 110) \cdot 44 =$ 25.52	$(64 / 110) \cdot 42 =$ 24.36	$(64 / 110) \cdot 24 =$ 13.92	
Total	44	42	24	110

This example is a little more complicated than the previous one because we have two variables: wealth and health status. This affects how we calculate our expected values, how we calculate the degrees of freedom, and how we test for significance.

If we are asking ourselves how wealth affects health, wealth is our independent variable and health is our dependent variable. Our null hypothesis will be that changing wealth does not affect health. So we look at the ratio of total sick to total people, and we expect that ratio to hold across the whole sick row because wealth should not change it. In our example, we have 46 sick people, of the 110 total: 46/110 = 0.42. Now we look at each of the cells in the sick row and fill in the expected counts. We have a total of 44 poor people, so we *expect* 42% of them to be sick. 42% of 44 is 18.48 (obviously we can't have .48 of a person, but that is OK). We do this for all cells in the row. The same principle holds for the healthy row. There are 64 healthy people of the 110 people sampled—64/110 = 0.58—so across all wealth categories we *expect* 58% of the people to be healthy.

Category	X_i	$E(X_i)$	$X_i - E(X_i)$	$(X_i - E(X_i))^2$	$(X_i - E(X_i))^2 /$ $E(X_i)$
Description	Observed counts	Expected counts	Difference of observed from expected	Difference of observed from expected squared	
Sick/Poor	20	18.48	1.52	2.31	.125
Sick/Middle	18	17.64	.36	.13	.001
Sick/Rich	8	10.08	-2.08	4.34	.43
Healthy/Poor	24	25.52	-1.52	2.31	.09

Healthy/Mid-dle	24	24.36	-.36	.13	.00053
Healthy/Rich	16	13.92	2.08	4.33	.31
Totals	110	110*	0**	-	X^2= .95653

* This total must always be the same as the observed total since it is the number of trials.
** This must always be zero, since if we observed more than we expected in one category, we must have observed less than we expected in others.

Don't forget to check to make sure we can use the chi-squared test. We do have at least 10 for each expected category count, but we also need to make sure we have at least two degrees of freedom. When the data have multiple variables, the degrees of freedom is given by:

$$df = (r - 1)(c - 1)$$

where r is the number of rows and c is the number of columns. The reason is that once we know rows $1...r - 1$ and once we know columns $1...c - 1$, we know the values for the cells in row r and column c.

So in our example we have $(2 - 1) \cdot (3 - 1) = 2$ degrees of freedom.

In this example we test for significance a little differently than the last. Here we use the shuffle method. The shuffle must preserve the marginals (i.e., the totals for rows and the totals for columns).

The row marginals are: total sick is 46 and total healthy is 64. The column marginals are: total poor is 44, total middle is 46, and total rich is 24. That is, you compute a chi-square value for this. Now a shuffle is a rearrangement that keeps the marginals the same. For example, if you move one from sick/poor to sick/middle, you would not affect the sick marginal but you would affect the poor and middle marginals, so you might move 1 from healthy/middle to healthy/poor. Another way to think about his is that we invent $20 + 18 + 8 + 24 + 24 + 16 = 110$ individuals, each associated with a health label and a wealth label. So the first 20 would be poor and sick, the next 18 middle and sick, etc. Then we shuffle the wealth labels and reevaluate the chi-squared. Clearly, the marginals don't change.

We can't change the marginals because when we test for significance we depend on our expected values and our expected values are computed from the marginals. Remember when we test for significance we simulate taking 10,000 samples — we use the expected probabilities to generate these samples.

4.3.5   Pseudocode & code

All code was tested using Python 2.3.

download code and input files

When we ran the ChiSquaredMulti.py code on the example above we got:

```
Observed chi-squared: 0.97
6241 out of 10000 experiments had a chi-squared greater than or equal to 0.97
```

```
Probability that chance alone gave us a chi-squared greater than or equal to 0.97
is 0.6241
```

Based on these results we conclude that wealth has no significant affect on health (remember that our observed values are made up!).

4.4 Fisher's Exact Test

4.4.1 Why and when

Fisher's exact test can be used instead of <u>chi-squared</u> when you have two variables (for example health and wealth), each having two categories (for example: sick, healthy and poor, rich), and one or more of the expected counts for the four possible categories ($2 \cdot 2 = 4$) are below 10 (remember: you cannot use chi-squared if even one expected count is less than 10). Like chi-squared, Fisher's exact test can be used to see if there is a relationship between the two variables as well as to measure the deviation of observed data from expectation.

Reputedly, Fisher was inspired to develop this test after an acquaintance, Dr. Muriel Bristol, claimed to be able to distinguish between a cup of tea in which the milk was poured before the tea and a cup in which the tea was poured before the milk[1]. In honor of Dr. Bristol, we will test this hypothesis.

4.4.2 Calculate with Example

Say we have collected the following data:

		Actual Order		Total
		Milk First	Tea First	
Claim of Tea Taster	Milk First	3	1	4
	Tea First	2	4	6
	Total	5	5	10

The above generalizes to:

		Variable A		Total
		Category A1	Category A2	
Variable B	Category B1	a	b	$a + b$
	Category B2	c	d	$c + d$
	Total	$a + c$	$b + d$	n

If our tea taster really can correctly identify which cups of tea had the milk poured first and which did not, then we expect a and d to be large (meaning he or she guessed correctly most of the time) and b and c to be small (incorrect guesses). If b and c are large, then our taster consistently misidentified the order of tea and milk. If our taster consistently misidentifies the order then he or she probably can distinguish between the two types, but has labeled them incorrectly. If our taster

---

1.   http://en.wikipedia.org/wiki/Muriel_Bristol

cannot distinguish between types of tea, then we expect a, b, c, and d to all be about the same, meaning the taster was wrong about as often as was correct.

Our null hypothesis is that our tea taster will not be able to tell the difference between the two types of teas. Notice here we are stating that we do not think that our taster can distinguish between the teas, we are *not* claiming that he or she cannot correctly identify them. So if our taster misidentified every single cup of tea, we would probably conclude that he or she can distinguish between the two types of tea. These are two different hypotheses and we will address the latter one later.

This is called an *exact* test because we know all of the possible 2 x 2 matrices we could have gotten while preserving the marginals (i.e., keeping the same row and column totals), and we know the exact probability of getting each matrix by chance given our null hypothesis is true. Remember that our p-value always includes the probability that chance alone gave us our observed data as well as the probability of chance alone giving us data even more *extreme* than ours. Being very clear about what question you are asking (what your null hypothesis is) will help you identify which cases are at least as extreme as your observed case.

The probability of getting matrix

a	b
c	d

is:

$$\frac{(a+b)!(c+d)!(a+c)!(b+d)!}{a!b!c!d!n!}$$

Remember that:

$$x! = x \cdot (x-1) \cdot (x-2) \cdot \ldots \cdot 1$$

And that $0! = 1$.

We know that the sum of the probabilities of all possible outcomes must be 1. So the denominator in the formula above represents the whole (i.e., all possible outcomes). The numerator in the above formula represents the number of times we expect to get this particular outcome. A large numerator yields a large probability, meaning this outcome is quite likely to occur. The numerator will be large when the matrix is balanced (i.e., a, b, c, and d are about equal). The more unbalanced the matrix is, the smaller the numerator will be. Bringing this back to our original tea problem, the greater the difference is between the number of teas our taster correctly identified (a and d) and the number he or she incorrectly identified (b and c), the smaller the likelihood that this difference is due to chance alone. This makes sense, if the taster gets them all mostly right, or all mostly wrong, it suggests that there is in fact a relationship of some kind between his or her decision making process, and the actual kind of tea he or she was given. If there was no relationship between the two variables (claim of the taster and actual order of milk and tea) then we expect our taster to be wrong about the same number of times as he or she is correct.

Here are all of the matrices we could have gotten given our row and column totals, including the one we did get, and their associated probabilities:

Matrix	Probability Calculation	Probability
0 4 **4** 5 1 **6** **5 5 10** (1 correct, 9 incorrect)	$((0 + 4)!(5 + 1)!(0 + 5)!(4 + 1)!) / (0!4!5!1!10!)$ $= (4!6!5!5!) / (0!4!5!1!10!)$ $= (6!5!) / (10!)$ $= (6·5·4·3·2·1·5·4·3·2·1) / (1·10·9·8·7·6·5·4·3·2·1)$ $= (5·4·3·2) / (10·9·8·7)$ $= (4·3) / (9·8·7)$ $= 1 / (3·2·7)$ $= 1 / 42$	0.0238
1 3 **4** 4 2 **6** **5 5 10** (3 correct, 7 incorrect)	$((1 + 3)!(4 + 2)!(1 + 4)!(3 + 2)!) / (1!3!4!2!10!)$ $= (4!6!5!5!) / (1!3!4!2!10!)$ $= (6·5·4·3·2·1·5·4·3·2·1·5·4·3·2·1) / (3·2·1·2·1·10·9·8·7·6·5·4·3·2·1)$ $= (5·4·5·4·3) / (10·9·8·7)$ $= 5 / 21$ $= 10 / 42$	0.2381
2 2 **4** 3 3 **6** **5 5 10** (5 correct, 5 incorrect)	$((2 + 2)!(3 + 3)!(2 + 3)!(2 + 3)!) / (2!2!3!3!10!)$ $= (4!6!5!5!) / (2!2!3!3!10!)$ $= (4·3·2·1·6·5·4·3·2·1·5·4·3·2·1·5·4·3·2·1) / (2·1·2·1·3·2·1·3·2·1·10·9·8·7·6·5·4·3·2·1)$ $= 20 / 42$	0.4762
observed frequency 3 1 **4** 2 4 **6** **5 5 10** (7 correct, 3 incorrect)	$((3 + 1)!(2 + 4)!(3 + 2)!(1 + 4)!) / (3!1!2!4!10!)$ $= 10 / 42$ (as above)	0.2381
4 0 **4** 1 5 **6** **5 5 10** (9 correct, 1 incorrect)	$((4 + 0)!(1 + 5)!(4 + 1)!(0 + 5)!) / (4!0!1!5!10!)$ $= 1 / 42$ (as above)	0.0238
The sum of all probabilities must be 1.		

Remember we are interested in matrices that are at least as extreme as the observed matrix. Our taster correctly identified 7 cups of tea and incorrectly identified 3. We decided that our null hypothesis is that our taster cannot distinguish between the two types of tea, which means that cases that are as extreme, or more extreme than our observed case, are those in which our taster got 3 or fewer incorrect, as well as those in which our taster got 3 or fewer correct. The first, second, fourth, and fifth matrices make up the set of matrices at least as extreme as the observed matrix (they have a probability that is less than or equal to our observed matrix). The sum of these probabilities gives us our p-value.

$$p\text{-value} = 0.0238 + 0.2381 + 0.2381 + 0.0238 = 0.5238$$

Assuming the null hypothesis is true and our tea taster cannot distinguish between cups of tea that had the milk poured in before the tea and those that did not, we would expect to get observed values such as ours 52% of the time. If the null hypothesis is true it is quite likely we would see results like ours, so we should not reject the null hypothesis; most likely our tea taster cannot differentiate between the two types of tea. Of course this does not mean that Dr. Bristol could not distinguish between the two, perhaps she had a more discriminating tea palate.

The above is an example of a two-tailed test. Now if our null hypothesis had been that our taster cannot correctly identify the tea cups, our definition of more extreme would only include cases in which our taster got even more correct than in our observed case. There is only one matrix in which that could have occurred: if she had gotten 9 correct and 1 incorrect. This is a one-tailed test. Our p-value here is the sum of the probability of getting matrix 4 and the probability of getting matrix 5.

$$p\text{-value} = 0.2381 + 0.0238 = 0.2619$$

So assuming our taster cannot correctly identify the cups of tea, we would expect to see an outcome such as ours about 26% of the time. This is still too probable for us to reject this hypothesis.

4.4.3   Pseudocode & code

All code was tested using Python 2.3.

download code and input files

Here is an example run of the FishersExactTestSig.py code:

```
Observed Fisher's Exact Test: 0.2619
3997 out of 10000 experiments had a Fisher's Exact Test less than or equal to
0.2619
Probability that chance alone gave us a Fisher's Exact Test of 0.2619 or less is
0.3997
```

4.5   One-Way ANOVA

4.5.1   Why and when

Given two or more groups, how would we measure how different they are? The mean of each group is a reasonable statistic to compare the groups. R. A. Fisher developed a statistic, called the *f*-statistic in his honor, for exactly this purpose.

4.5.2   Calculate with example

The one-way ANOVA, or single-factor ANOVA, tests the effect of a single independent variable on the groups. The groups must be categorized on this single variable. The idea is that we want to compare the variance between the groups with the variance within the groups. If the variance between the groups is significantly greater than the variance within the groups, we conclude that the independent variable does affect the groups.

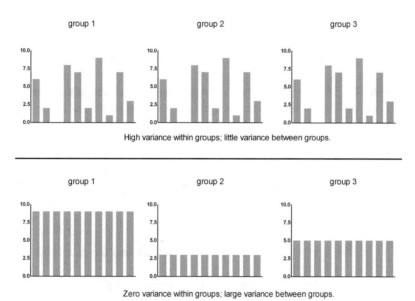

Figure 4.1:   The difference between within group variance and between group variance.

Here we have an exaggerated example of the difference between within group variance and between group variance. In the top row we have three groups with large within group variance, but that are exactly the same so there is no between group variance. In the bottom row we have three groups that have no within group variance, but the groups are different from each other, and therefore have a large between group variance.

Note: The number of elements in each group need not be equal.

How to compute the f-statistic:

This algorithm involves a lot of steps so we will explain it using a simple example and a table. We have three different drugs used as treatment. We want to test the effect of our independent variable, the drug given as treatment, on the number of days it takes a patient to fully recover, so we separate patients into three groups based on the drug they have been given and record the number of days it takes for each to recover:

$$\text{Drug A} = 45, 44, 34, 33, 45, 46, 34$$

$$\text{Drug B} = 34, 34, 50, 49, 48, 39, 45$$
$$\text{Drug C} = 24, 34, 23, 25, 36, 28, 33, 29$$

We compare the between group variance and the within group variance using the ratio:

$$MS_{Between}/MS_{Within}$$

where $MS_{Between}$ is the between group variance, referred to as the between group mean squares and MS_{Within} is the within group variance, referred to as the within group mean squares.

To compute the within group mean squares we compute the within group sum of squares and divide it by the within group degrees of freedom. The within group sum of squares will be the sum of each group's sum of squares. This may sound confusing, but it is actually straightforward. To get one group's sum of squares we compute the difference between each value in the group and that value's own group mean, square the result, and then sum these squares. W_{SS} is the sum of the sum of squares for all groups.

$$W_{SS} = \sum_{g=1}^{k} \sum_{i=1}^{n_g} (x_i - \bar{X}_g)^2$$

where k is the number of groups, n_g is the number of elements in group g, x_i is the ith value in the group g, and $\bar{X}_g$ is the group mean.

Here we compute the sum of squares for each group in our example:

$$\text{Drug A}_{SS} = (45 - 40.14)^2 + (44 - 40.14)^2 + (34 - 40.14)^2 + (33 - 40.14)^2 + (45 - 40.14)^2 + (46 - 40.14)^2 + (34 - 40.14)^2 = 222.86$$

$$\text{Drug B}_{SS} = (34 - 42.71)^2 + (34 - 42.71)^2 + (50 - 42.71)^2 + (49 - 42.71)^2 + (48 - 42.71)^2 + (39 - 42.71)^2 + (45 - 42.71)^2 = 291.43$$

$$\text{Drug C}_{SS} = (24 - 29)^2 + (34 - 29)^2 + (23 - 29)^2 + (25 - 29)^2 + (36 - 29)^2 + (28 - 29)^2 + (33 - 29)^2 + (29 - 29)^2 = 168$$

The within group sum of squares equals 682.29; it is the sum of the above group sum of squares.

The last piece of information we need to compute the within group mean squares (the within group variance) is the degrees of freedom for the within group sum of squares. In general,

$$df = n - 1$$

where n is equal to the number of values in the group.

The degrees of freedom measures the number of independent pieces of information on which our statistic is based. When computing the group df we subtract one from the group count because if we know the mean and n, once we know the $(n - 1)^{\text{th}}$ values we know what the nth value must be.

Therefore, the degrees of freedom for the within group sum of squares is:

$$W_{df} = N - k$$

where N is equal to the total number of values in all the groups, and k is equal to the number of groups. We subtract k from N because for each group we lose one degree of freedom.

Computing the between group mean squares is very similar to computing the within group mean squares, except that for each value we get the difference between the value's own group mean and the total mean (computed by pooling all the values).

$$B_{SS} = \sum_{i=1}^{N} (\bar{X}_g - \bar{X})^2$$

Where N is equal to the total number of values in all the groups, $\bar{X}_g$ is the group mean of the ith value in the pool of all values, and is the overall mean. Because every value in a particular group will have the same (group mean – total mean), the above formula can be simplified to:

$$B_{SS} = \sum_{g=1}^{k} n_g (\bar{X}_g - \bar{X})^2$$

Where k is equal to the number of groups, n_g is equal to the total number of values in group g, $\bar{X}_g$ is the group mean and $\bar{X}$ is the overall mean. That is, B_{SS} multiplies the squared difference of the group mean and total mean by the number of values in the group.

In our example:

$$B_{SS} = 7 \cdot (40.14 - 36.91)^2 + 7 \cdot (42.71 - 36.91)^2 + 8 \cdot (29 - 36.91)^2 = 812.65$$

The between group degrees of freedom is:

$$B_{df} = k - 1$$

where k is equal to the number of groups. We subtract 1 from k because if we know the number of elements in each group and the total group mean, once we know the $(k - 1)^{\text{th}}$ group means, we know what the kth group mean must be.

Group	Description	A	B	C
Count	Number of values in group	7	7	8
$\bar{X}_g$	Within group mean, where g is the group number	40.14	42.71	29.0

G_{ss}	Sum of squares for each group	222.86	291.43	168
W_{ss}	Within group sum of squares	682.29		
G_{df}	Group degrees of freedom	6	6	7
W_{df}	Within group degrees of freedom	19		
$\overline{X}$	Total mean, the mean of all the values pooled	36.91		
B_{ss}	Between group sum of squares	812.65		
B_{df}	Between group degrees of freedom	2		
T_{ss}	Total sum of squares	1491.81		
MS_W	Within group variance (mean squares)	35.91		
MS_B	Between group variance (mean squares)	406.33		
f	Measure of how much bigger the between group variance is than the within group variance	11.31		

The total sum of squares measures the variance of all the values. $T_{SS} = W_{SS} + B_{SS}$. Therefore $B_{SS} = T_{SS} - W_{SS}$ and $W_{SS} = T_{SS} - B_{SS}$ hold. If it is easier for you to compute W_{SS} and T_{SS}, you can use these to compute B_{SS}.

$$T_{SS} = \sum_{i=1}^{N} (x_i - \overline{X})^2$$

where N is equal to the total number of values in all the groups, x_i is the ith value in the pool of all values, and $\overline{X}$ is the overall mean.

We now have our f-statistic, 11.31, so we know that the between group variance is greater than the within group variance, but we still do not know whether or not this difference is **significant**. We use the shuffle method to compute the significance of our f-statistic.

4.5.3   Pseudocode & code

All code was tested using Python 2.3.

<u>**download code and input files**</u>

When we ran the OneWayAnovaSig.py code on the example above we got:

```
Observed F-statistic: 11.27
10 out of 10000 experiments had a F-statistic difference greater than or equal to
11.27
Probability that chance alone gave us a F-statistic of 11.27 or more is 0.001
```

OK, the difference may be significant, but is it large? Does it matter?

All code was tested using Python 2.3.

<u>**download code and input files**</u>

When we ran the OneWayAnovaConf.py code we got:

```
Observed F-statistic: 11.27
We have 90.0 % confidence that the true F-statistic is between: 6.28 and 29.60
***** Bias Corrected Confidence Interval *****
We have 90.0 % confidence that the true F-statistic is between: 4.47 and 20.35
```

All f tells you is whether or not there is a <u>significant</u> difference between the group means, not which group means are significantly different from each other. You can do follow-on tests to find out which group means are different since we have decided our f-statistic indicates the group means significantly differ. You can now compare the different groups by <u>comparing the group means</u>. You could do this pairwise comparison between every group, or if you have a control group you can do pairwise comparisons between the control and every other group.

4.6   Multi-way ANOVA

4.6.1   Why and when

We use a multi-way ANOVA, or multi-factor ANOVA, when we want to test the influence of two or more *factors (independent variables)*, on our outcome. Each factor must be discrete, or categorical. For example, this is not the appropriate test to ask if sex and weight effect some measure of health, unless weight is divided into categories such as underweight, healthy weight, etc.

In our <u>one-way ANOVA</u> example, we tested the effect of only one factor—the drug a patient was given—on outcome. We used this factor to split our outcome measures into different groups, and then used the f-statistic to decide whether or not our groups were significantly different. The steps we take for the multi-way ANOVA are very similar to the <u>one-way ANOVA</u>, but with each additional factor we add additional sources of variance, and we must take these additional sources of variance into consideration.

4.6.2   Calculate with example

For demonstrative purposes we will do a two-factor ANOVA comparing drug treatment and sex to outcome.

Here is a table that summarizes our data.

factor D		factor S		Totals
		Category s_1	Category s_2	
factor D	Category d_1	15 12 13 16 $\overline{X}_{d_1 s_1} = 14$	13 13 12 11 $\overline{X}_{d_1 s_2} = 12.25$	$\overline{X}_{d_1} = 13.13$
	Category d_2	19 17 16 15 $\overline{X}_{d_2 s_1} = 16.75$	13 11 11 17 $\overline{X}_{d_2 s_2} = 13$	$\overline{X}_{d_2} = 14.88$
	Category d_3	14 13 12 17 $\overline{X}_{d_3 s_1} = 14$	11 12 10 $\overline{X}_{d_2 s_3} = 11$	$\overline{X}_{d_3} = 12.71$
Totals		$\overline{X}_{s_1} = 14.92$	$\overline{X}_{s_2} = 12.18$	$\overline{X} = 13.6$

We have split our outcome data into six groups. In a multi-way ANOVA if we have F factors, we end up with

$$\prod_{f=1}^{F} n_f$$

groups, where n_f is the number of categories in factor f. In our example, we have factor D, drug, of which there are 3 possibilities, and factor S, sex, of which there are 2 possibilities. Since there are 3 categories in factor D and 2 in factor S, we end up with $3 \cdot 2 = 6$ groups.

In the one-way ANOVA we compared the between group variance to the within group variance to see if there was a difference in the means of our groups. In the one-way ANOVA we only had one factor that could have been introducing variance outside of natural variance between the participants in the study. In a two-factor ANOVA, we have:

1.   the natural variance between the study participants

2.   the variance in the groups caused by factor D

3.   the variance in the groups caused by factor S

4.   the variance in the groups caused by the *interaction* of factor D and factor S.

Using a two-factor ANOVA we can find out:

1. how factor D influences the group means
2. how factor S influences the group means
3. how the *interaction* of factor D and factor S influences the group means

Remember: The sum of squares is the sum of squared differences of each value from some mean. We use the sum of squares and the degrees of freedom to measure variance in a group or between groups. So here we will use this to measure all of the sources of variance.

We compute the within group variation just as we did in the <u>one-way ANOVA</u>.

$$W_{SS} = \sum_{g=1}^{k} \sum_{i=1}^{n_g} (x_i - \bar{X}_g)^2$$

where k is the number of groups, n_g is the number of elements in group g, x_i is the ith value in the group g, and $\bar{X}_g$ is the group mean.

Here we compute the sum of squares for each group in our example:

Group $d_1 s_{1SS} = (15 - 14)^2 + (12 - 14)^2 + (13 - 14)^2 + (16 - 14)^2 = 10$

Group $d_1 s_{2SS} = (13 - 12.25)^2 + (13 - 12.25)^2 + (12 - 12.25)^2 + (11 - 12.25)^2 = 2.75$

Group $d_2 s_{1SS} = (19 - 16.75)^2 + (17 - 16.75)^2 + (16 - 16.75)^2 + (15 - 16.75)^2 = 8.75$

Group $d_2 s_{2SS} = (13 - 13)^2 + (11 - 13)^2 + (11 - 13)^2 + (17 - 13)^2 = 24$

Group $d_3 s_{1SS} = (14 - 14)^2 + (13 - 14)^2 + (12 - 14)^2 + (17 - 14)^2 = 14$

Group $d_3 s_{2SS} = (11 - 11)^2 + (12 - 11)^2 + (10 - 11)^2 = 2$

Within Group Sum of Squares = 10 + 2.75 + 8.75 + 24 + 14 + 2 = 61.5

The degrees of freedom for the within group sum of squares is:

$$W_{df} = N - k$$

where N is equal to the total number of values in all the groups, and k is equal to the number of groups. We subtract k from N because for each group we lose one degree of freedom.

In our example, the degrees of freedom for the within group sum of squares is 17 because we have 23 values in total and 6 groups.

Now we can compute the within group mean squares, which is the within group sum of squares divided by the within group degrees of freedom.

$$W_{MS} = 61.5 / 17 = 3.62$$

The between group sum of squares will include the variance from 3 sources: factor D, factor S, and the interaction of factors D and S. So we will refer to the between group sum of squares as the total between group sum of squares, to indicate that it incorporates all of these sources of variance. It is computed just as our between group sum of squares in the one-way ANOVA.

$$B_{SS} = \sum_{i=1}^{N} (\bar{X}_g - \bar{X})^2$$

where N is equal to the total number of values in all the groups, $\bar{X}_g$ is the group mean of the ith value in the pool of all values, and $\bar{X}$ is the overall mean.

In our example the total between group sum of squares equals:

$$B_{SS} = 4 \cdot (14 - 13.6)^2 + 4 \cdot (12.25 - 13.6)^2 + 4 \cdot (16.75 - 13.6)^2 + 4 \cdot (13 - 13.6)^2 + 4 \cdot$$
$$(14 - 13.6)^2 + 4 \cdot (11 - 13.6)^2 = 69.98$$

The between group degrees of freedom is:

$$B_{df} = k - 1$$

where k is equal to the number of groups. We subtract 1 from k because if we know the number of elements in each group and the total group mean, once we know the $(k - 1)^{th}$ group means, we know what the kth group mean must be.

In our example, the between group degrees of freedom is 5, because we have 6 groups.

Now we want to split B_{SS} into its 3 parts: the main effects of factor D, the main effects of factor S, and the interaction effect.

To figure out how factor D influences the group means, we just do a one-way ANOVA where our groups are d_1, d_2, and d_3.

$$factorD_{SS} = \sum_{i=1}^{N} (\bar{X}_d - \bar{X})^2$$

where N is equal to the total number of values in all the groups, i is the current value, d is the current group of factor D, $\bar{X}_d$ is the mean of the values in group d, and $\bar{X}$ is the overall mean. Because every value in a particular group d will have the same (group mean - total mean), the above formula can be simplified to:

$$factorD_{SS} = \sum_{d=1}^{D} n_d (\bar{X}_d - \bar{X})^2$$

where D is equal to the number of groups, or categories, in factor D, d is the current group of factor D, n_d is equal to the total number of values in group d, $\bar{X}_d$ is the mean of the values in group d, and $\bar{X}$ is the overall mean. We just multiplied the squared difference of the group mean and total mean by the number of values in the group.

In our example:

$$\text{factor D}_{SS} = 8 \cdot (13.13 - 13.6)^2 + 8 \cdot (14.88 - 13.6)^2 + 7 \cdot (12.71 - 13.6)^2 = 42.92$$

$$\text{factorD}_{df} = D - 1$$

In our example factor D has 2 degrees of freedom.

Now we can compute factor D mean squares, which is the factor D sum of squares divided by the factor D degrees of freedom.

$$\text{factor D}_{MS} = 42.92 / 2 = 21.46$$

We compute how factor S influences the group means in the same way.

$$\text{factorS}_{SS} = \sum_{i=1}^{N} (\bar{X}_s - \bar{X})^2$$

or

$$\text{factorS}_{SS} = \sum_{s=1}^{S} n_s (\bar{X}_s - \bar{X})^2$$

In our example:

$$\text{factor S}_{SS} = 12 \cdot (14.92 - 13.6)^2 + 11 \cdot (12.18 - 13.6)^2 = 20.3$$

and

$$\text{factorS}_{df} = S - 1$$

In our example factor S has 1 degree of freedom.
Now we can compute factor S mean squares, which is the factor S sum of squares divided by the factor S degrees of freedom.

$$\text{factor S}_{MS} = 20.3 / 1 = 20.3$$

Finally,

$$\text{interaction}_{SS} = B_{SS} - \text{factorD}_{SS} - \text{factorS}_{SS}$$

In our example:

$$\text{interaction}_{SS} = 69.98 - 42.92 - 20.3 = 6.76$$

and

$$\text{interaction}_{df} = B_{df} - \text{factorD}_{df} - \text{factorS}_{df}$$

$$\text{interaction}_{df} = 5 - 2 - 1 = 2$$

Now we can compute the interaction mean squares, which is the interaction sum of squares divided by the interaction degrees of freedom.

$$interaction_{MS} = 6.76 / 2 = 3.38$$

Remember from the <u>one-way ANOVA</u> that our f-statistic compares the between group variance and the within group variance. The same holds with the multi-way ANOVA, except we will compare each part of the between group variance with the within group variance, to answer each of three hypotheses. So instead of:

$$\frac{MS_{Between}}{MS_{Within}}$$

we will compare factor D mean squares with the within group mean squares, factor S mean squares with the within group mean squares, and the interaction mean squares with the within group mean squares. In other words, we are comparing the variance caused by factor D with the variance within the groups, the variance caused by factor S with the variance within the groups, and the variance caused by the interaction of factors D and S to the variance.

$$\frac{MS_{factorD}}{MS_{Within}}, \frac{MS_{factorS}}{MS_{Within}} \text{ and } \frac{MS_{interaction}}{MS_{Within}}$$

Here we summarize our results:

	SS	df	MS	f
Within	61.5	17	3.62	
Between	69.98	5		
factor D	42.92	2	21.46	21.46/3.62 = 5.93
factor S	20.3	1	20.3	20.3/3.62 = 5.61
Interaction of D and S	6.73	2	3.38	3.38/3.62 = 0.93

We still don't know if this is significant.

4.6.3   Pseudocode & code

All code was tested using Python 2.3.

<u>download code and input files</u>

When we ran the TwoWayAnovaSig.py code on the example above we got:

```
Observed F-statistic: 0.93
4152 out of 10000 experiments had a F-statistic greater than or equal to 0.93
Probability that chance alone gave us a F-statistic of 0.93 or more is 0.4152
```

OK, the difference *is not* <u>significant</u>. If it were we would check if it is large; if the difference matters.

All code was tested using Python 2.3.

<u>download code and input files</u>

When we ran the TwoWayAnovaConf.py code we got:

```
Observed F-statistic: 0.93
We have 90.0 % confidence that the true F-statistic is between: 0.50 and 8.44
***** Bias Corrected Confidence Interval *****
We have 90.0 % confidence that the true F-statistic is between: 0.20 and 2.04
```

4.7 Linear Regression

4.7.1 Why and when

Linear regression is closely tied to <u>linear correlation</u>. Here we try to find the line that best fits our data so that we can use it to predict y given a new x. Correlation measures how tightly our data fits that line, and therefore how good we expect our prediction to be.

4.7.2 Calculate with example

> BE CAREFUL
> Regression can be misleading when there are outliers or a nonlinear relationship.

The first step is to draw a scatter plot of your data. Traditionally, the independent variable is placed along the x-axis, and the dependent variable is placed along the y-axis. The dependent variable is the one that we expect to change when the independent one changes. Note whether the data looks as if it lies approximately along a straight line. If it has some other pattern, for example, a curve, then a linear regression should not be used.

The most common method used to find a regression line is the least squares method. Remember that the equation of a line is:

$$y = bx + a$$

where b equals the slope of the line and a is where the line crosses the y-axis.

The least squares method minimizes the vertical distances between our data points (the observed values) and our line (the predicted values).

The equation of the line we are trying to find is:

$$y' = bx + a$$

where y' is the predicted value of y for some x.

We have to calculate b and a.

$$b = \frac{XY_{SP}}{X_{SS}}$$

where XY_{SP} is the sum of products:

$$XY_{SP} = \sum_{i=1}^{N} (x_i - \bar{X})(y_i - \bar{Y})$$

and X_{SS} is the sum of squares for X:

$$X_{SS} = \sum_{i=1}^{N} (x_i - \bar{X})^2$$

In our example:

	x_i	y_i	$\bar{X}$	$\bar{Y}$	$(x_i - \bar{X})$	$(y_i - \bar{Y})$	$(x_i - \bar{X})(y_i - \bar{Y})$	$(x_i - \bar{X})^2$
	1350	3.6	1353	3.5	-3	.1	-.3	9
	1510	3.8	1353	3.5	157	.3	47.1	24649
	1420	3.7	1353	3.5	67	.2	13.4	4489
	1210	3.3	1353	3.5	-143	-.2	28.6	20449
	1250	3.9	1353	3.5	-103	.4	-41.2	10609
	1300	3.4	1353	3.5	-53	-.1	5.3	2809
	1580	3.8	1353	3.5	227	.3	68.1	51529
	1310	3.7	1353	3.5	-43	.2	-8.6	1849
	1290	3.5	1353	3.5	-63	0	0	3969
	1320	3.4	1353	3.5	-33	-.1	3.3	1089
	1490	3.8	1353	3.5	137	.3	41.1	18769
	1200	3.0	1353	3.5	-153	-.5	76.5	23409
	1360	3.1	1353	3.5	7	-.4	-2.8	49
Totals	-	-	-	-	-	-	230.5	163677

$XY_{SP} = 230.5$ and $X_{SS} = 163677$ and $230.5 / 163677 = 0.0014$, so $b = 0.0014$.

The regression line will always pass through the point $(\bar{X}, \bar{Y})$ so we can plug this point into our equation to get a, where the line passes through the y-axis.

So far we have: $y' = 0.0014x + a$, so to solve for a we rearrange and get: $a = y' - 0.0014x$. Plug in (1353, 3.5) for (y', x) and we get $a = 3.5 - (0.0014 \cdot 1353) = 1.6058$. Our final regression equation is:

$$y' = .0014x + 1.6058$$

We shuffle the y values to break any relationship that exists between x and y. We do this 10,000 times, computing the slope each time to get the probability of a slope greater than or equal to ours given no relationship between x and y. If the slope was negative we would test to see the probability of getting a slope less than or equal to ours given no relationship between x and y. Remember that if the slope is zero, then x is not related to an increase or decrease in y.

4.7.3   Pseudocode & code

All code was tested using Python 2.3.

download code and input files

When we ran the RegressionConf.py code on the example above we got:

```
Line of best fit for observed data:  y' = 0.0014 x +  1.6333
165 out of 10000 experiments had a slope greater than or equal to 0.0014 .
The chance of getting a slope greater than or equal to 0.0014 is 0.0165 .
```

OK, the difference may be significant, but is it large? Does it matter?

download code and input files

When we ran the RegressionConf.py code on the example above we got:
```
Line of best fit for observed data:  y' = 0.0014 x +  1.6333
We have 90.0 % confidence that the true slope is between: 0.0005 and 0.0022
***** Bias Corrected Confidence Interval *****
We have 90.0 % confidence that the true slope is between: 0.0004 and 0.0022
```

4.8   Linear Correlation

4.8.1   Why and when

The correlation coefficient is a measure of strength and direction of a linear relationship between two random variables. Remember, we used regression to define the linear relationship between the two variables, and now we want to know how well X predicts Y.

4.8.2   Calculate & example

Draw scatter plot of data. Place the independent variable along the x-axis, and the dependent variable along the y-axis. We want to see what happens to the dependent variable when the independent one changes. Note whether the data looks as if it lies approximately along a straight line. If it has some other pattern, for example, a curve, then the linear correlation coefficient should not be used.

The correlation coefficient is always between -1 and 1. Values close to 1 indicate a strong positive association, meaning that as X increases we expect Y to increase (positive sloping line). Values close to -1 indicate a strong negative association, meaning that as X increases we expect Y to decrease (negative sloping line). A value of zero indicates that there is no relationship between the variables, that is, that knowing X does not help you predict Y.

The correlation coefficient, r, is computed by comparing the observed covariance (a measure of how much X and Y vary together) to the maximum possible positive covariance of X and Y.:

$$r = \frac{XY_{SP}}{\sqrt{X_{SS}Y_{SS}}}$$

where XY_{SP} is the sum of products, X_{SS} is the sum of squares for X, and Y_{SS} is the sum of squares for Y:

$$XY_{SP} = \sum_{i=1}^{N}(x_i - \bar{X})(y_i - \bar{Y})$$

$$X_{SS} = \sum_{i=1}^{N}(x_i - \bar{X})^2$$

$$Y_{SS} = \sum_{i=1}^{N}(y_i - \bar{Y})^2$$

Let us look closely at XY_{SP}, the numerator in our equation. First, notice that neither $(x_i - \bar{X})$ nor $(y_i - \bar{Y})$ is squared. So both could be positive, both could be negative, or one could be positive and one could be negative. This means the product of these can be either negative or positive. Now let us think about a few scenarios. Suppose there is a positive relationship between X and Y, so in general as X increases Y increases. That means in general we would expect that while we are looking at Xs below the mean of all Xs, we also expect the Ys we look at to be below the mean of all Ys. So we would get a negative times a negative resulting in a positive number. We would also expect that when we examine Xs above the mean of all Xs, we would see Ys that are also above the mean of all Ys. So we would get a positive times a positive, resulting in yet another positive number. In the end we should get a sum of mostly positive numbers. If you go through the same steps for a hypothetical negative relationship you would end up summing mostly negative numbers, resulting in a relatively large negative number. The numerator gives us our positive or negative association. If there is no relationship between X and Y, we expect get some negative products and some positive products, the sum of which will cancel many of these values out.

In the denominator, $(x_i - \bar{X})$ and $(y_i - \bar{Y})$ are both squared before being multiplied together— resulting in a positive number. This means the denominator is always positive, and $(x_i - \bar{X})$ and $(y_i - \bar{Y})$ cannot cancel each other out when these products are summed.

If X and Y have a strong relationship, the absolute value of the observed covariance will be close to the maximum possible positive covariance, yielding a r close to 1 or -1.

Compute correlation coefficient:

	x_i	y_i	$\bar{X}$	$\bar{Y}$	$(x_i - \bar{X})$	$(y_i - \bar{Y})$	$(x_i - \bar{X})(y_i - \bar{Y})$	$(x_i - \bar{X})^2$	$(y_i - \bar{Y})^2$
	1350	3.6	1353	3.5	-3	.1	-.3	9	.01
	1510	3.8	1353	3.5	157	.3	47.1	24649	0.09
	1420	3.7	1353	3.5	67	.2	13.4	4489	0.04
	1210	3.3	1353	3.5	-143	-.2	28.6	20449	0.04
	1250	3.9	1353	3.5	-103	.4	-41.2	10609	.16
	1300	3.4	1353	3.5	-53	-.1	5.3	2809	0.01
	1580	3.8	1353	3.5	227	.3	68.1	51529	.09
	1310	3.7	1353	3.5	-43	.2	-8.6	1849	0.04
	1290	3.5	1353	3.5	-63	0	0	3969	0
	1320	3.4	1353	3.5	-33	-.1	3.3	1089	0.01
	1490	3.8	1353	3.5	137	.3	41.1	18769	0.09
	1200	3.0	1353	3.5	-153	-.5	76.5	23409	.25
	1360	3.1	1353	3.5	7	-.4	-2.8	49	0.16
Totals	-	-	-	-	-	-	230.5	163677	0.99

$163677 \cdot .99 = 162040.23$, the square root of 162040.23 is 402.54, $230.5 / 402.54 = 0.57$, so $r = 0.57$.

We still don't know if r is significant. The null hypothesis is that there is no relationship between X and Y. We test this by using the shuffle method to break any relationship that may exist between the two, and see what the chances are that we get an r greater than or equal to .57.

4.8.3   Pseudocode & code

All code was tested using Python 2.3.

<u>download code and input files</u>

When we ran the CorrelationSig.py code on the example above we got:

```
Observed r: 0.57
151 out of 10000 experiments had a r greater than or equal to 0.57
Probability that chance alone gave us a r greater than or equal to 0.57 is
0.02
```

All code was tested using Python 2.3.

<u>download code and input files</u>

When we ran the CorrelationConf.py code on the example above we got:

```
Observed r: 0.58
We have  90.0 % confidence that the true r is between: 0.50 and 0.65
***** Bias Corrected Confidence Interval *****
We have 90.0 % confidence that the true r is between: 0.49 and 0.64
```

Correlation does not imply causation. Something other than X (but related to X) could be causing the changes in Y. So we can use X to predict Y, but a change in X does not necessarily cause a change in Y. A classic example is that, before the polio vaccine, there was a positive correlation between sales of soda and the outbreak of polio. This doesn't mean that soda caused polio. So what was going on? It turned out that polio spread more easily in the summer when many people played together at pools and on the beach. That was also the time of year when soda sales went up. It is important to be careful when making assumptions about what causes what.

4.9   Multiple Regression

See the <u>CART</u> algorithm.

4.10   Multiple Testing

4.10.1 Why and when

Any time you measure the behavior of multiple entities (e.g., genes, plants, or other experimental subjects) e1, ...e$_n$, in an experiment, there is a chance that you will get a behavior that differs substantially from what is expected, at least for some of the entities. For example, if you do 10,000 experiments, and the probability of getting the unusual behavior is even as small as 0.05%, then you would actually expect to get about 5 entities displaying the unusual behavior. When you do a large number of tests, you run the risk of deciding the unusual behavior you come across is significant, when in fact it is expected. So how do you deal with this problem? There are several ways to handle it, and which you choose depends partly on whether it is more important to for you to be precise (few false positives) or have high recall (few false negatives). Use the Bonferroni correction to avoid false negatives (see the Family Wise Error Rate section below).

4.10.2 Family Wise Error Rate

The family wise error rate (FWER) is the probability of getting a <u>type I error</u> (deciding that the null hypothesis is false when it is true) at least once when doing multiple significance tests.

Remember that the *p*-value x of an individual test result is defined as the probability that when the null hypothesis is true (i.e. the treatment has no effect), then this test individual test will show a value (e.g. a difference in the means) at least as great as the one observed with probability x. The key phrase is "individual test". But what if there are a family of tests?

Use the Bonferroni correction to get a family-wise error rate less than *q*, by rejecting a null hypothesis only if the *p*-value multiplied by the number of experiments (significance tests run), is less than *q*.

By doing this, we have make it much harder for any individual significance test to reach significance. Say we want our family wise error rate to be less than 0.05 and we are running 10,000 tests (a number of tests not uncommon in biology). If we have a *p*-value of 0.0002, $(0.0002 \cdot 10{,}000) = 2$, and 2 is not less than 0.05, so we cannot reject the null hypothesis even though we have a very small *p*-value. This is a much more stringent test for significance. By decreasing the likelihood of <u>type I errors</u>, we increased the likelihood of <u>type II errors</u>, so we make it more likely that for an individual test we do not reject the null hypothesis when in fact we should.

4.10.3 False Discovery Rate

The false discovery rate (FDR) controls the expected proportion of type I errors when doing multiple significance tests. Frequently it is preferable to have some false positives (type I errors) in the results than to have false negatives (type II errors). For example, say you have an experiment in which you want to find the genes most likely to be active in breast cancer. So you are looking at tens of thousands of genes in maybe 100 patients. It is probably better to include a few genes that are not actually involved in breast cancer, than to exclude some that are. Using the false discovery rate you can say "here are the genes we predict are involved in breast cancer, with a FDR of 0.3." So if you have selected 10 genes, you expect 3 of them to be false positives.

The first step of the approach given by Benjamini and Hochberg is to order the *p*-values of all the experiments:

$$P_{(1)} \leq \ldots \leq P_{(m)}$$

where *m* is the total number of experiments. Notice that the first *p*-value is the smallest, and therefore the most significant.

The number of experiments where we reject the null hypothesis is given by:

$$t = \max\left\{ i : P(i) \leq q\left(\frac{i}{m}\right) \right\}$$

where q is the target false discovery proportion and m is the total number of significance tests. i is the number of experiements where we reject the null hypothesis. The formula can be read as: our cutoff t is the maximum i where $P(i)$ is less than or equal to $(i \,/\, m)q$.

Suppose we set q to be 0.05 and we get some value t. We will reject the null hypothesis for all tests (e.g. genes in a biology experiment) i from 1 to t, asserting that those t genes have been significantly affected by the treatment. We can assert further that the resulting false discovery estimate among these t genes is 0.05.

The justification for the assertion of the false discovery rate is as follows: if all m test differences were in fact the result of chance and the null hypothesis held for them, then we'd expect the p-values to be uniformly distributed. That is, there would be as many test results having p-values between, say, 0% and 3% as between 55% and 58%. That holds because of the definition of p-value that we mentioned before. So, up to p-value $P(t)$ we'd expect a number $P(t)m$ false positives out of a total of t genes that have been asserted to be positive.

This gives us a false discovery rate equal to the number of false positives divided by the number declared to be positive = $(P(t)m)/t$. Rewriting the above equation for t we get that $(m/t)P(t) \leq q$ or equivalently $(P(t)m)/t \leq q$.

So, the resulting q properly estimates the false discovery rate under the slightly conservative assumption that there will be $P(t)m$ false positives up to a p-value of $P(t)$. A less conservative but somewhat more complex approach has been suggested by John Storey and Robert Tibshirani. You can find slides explaining that approach at **http://cs.nyu.edu/cs/faculty/shasha/papers/Qvalseasy.ppt**.

Chapter 4 Exercises

The data sets you need for these exercises will be at **http://cs.nyu.edu/cs/faculty/shasha/papers/stateasyExerciseData.zip**

Throughout these exercises, there are two running data sets: (i) one concerning drug vs. placebos with varying dosages and (ii) one concerning gene responses to a certain pair of inputs A and B (that is each gene was tested 4 times in 25 samples for each of 4 input types: no input, A alone, B alone, A and B together).

In the drug data set, higher response values are a good thing. The dosages are: 100mg/m2, 75mg/m2, 50mg/m2, placebo

For the gene test, let there be 20 genes, each replicated 4 times in a total of 100 samples.
Gene no input, A alone, B alone, AB together (each gene 4 times)

1.   For the drug case, find the confidence interval of the mean for the 75mg/m2 dose.

2. For the smallest dose drug vs. placebo case determine whether the difference is significant and the confidence interval of the difference.

3. For the drug case, try to find the influence of drug dosage on result.

    Hint: do one-way ANOVA on dosage.

4. Use two-way anova to test the influence of factors A and B on the gene expressions. Each factor is divided into categories given and not given. Use all 20 genes.

5. For the gene case, determine whether A and B together have a significantly different mean from no input, from A alone, from B alone. Compute the p-value for each gene for each case.

    AB vs. No input
    AB vs. A
    AB vs. B

    20 genes • 3 tests = 60

6. Get a Bonferroni corrected p-value using the results from problem 5. Use a Bonferroni cutoff of 0.05.

7. The Bonferroni correction is a very strict multiple testing correction. By using it you are choosing to increase the number of false negatives rather than have false positives in your results. If you prefer to be more inclusive, and have more false positives you would use a more permissive multiple testing correction such as the Hochberg and Benjamini correction. Use that method with a False Discovery Rate of 10% instead of using the Bonferroni correction and see which genes you get for the genes of question 5.

Chapter 4 Solutions

The data sets you need for these exercises will be at **http://cs.nyu.edu/cs/faculty/shasha/papers/ stateasyExerciseData.zip**

Throughout these exercises, there are two running data sets: (i) one concerning drug vs. placebos with varying dosages and (ii) one concerning gene responses to a certain pair of inputs A and B (that is each gene was tested 4 times in 25 samples for each of 4 input types: no input, A alone, B alone, A and B together).

In the drug data set, higher response values are a good thing. The dosages are: 100mg/m2, 75mg/ m2, 50mg/m2, placebo

For the gene test, let there be 20 genes, each replicated 4 times in a total of 100 samples.
Gene no input, A alone, B alone, AB together (each gene 4 times)

2. For the smallest dose drug vs. placebo case determine whether the difference is significant and the confidence interval of the difference.

Solution: When we ran Diff2MeanSig.py with the input file Diff2MeanCh4Ex2.vals we got:
Observed difference of two means: 5.27
2740 out of 10000 experiments had a difference of two means greater than or equal to 5.27 .
The chance of getting a difference of two means greater than or equal to 5.27 is 0.274 .

We cannot reject the null hypothesis here. It is too likely that the difference between the 50mg/m2 dose and the placebo is due to chance alone.

Next we ran Diff2MeanConf.py with the same input file and got:

Observed difference between the means: 5.27
We have 90.0 % confidence that the true difference between the means is between: -8.33 and 18.87

Here we can see that the 50mg/m2 dose should be better than the placebo in some cases and the placebo to be better in others.

4.   Use two-way anova to test the influence of factors A and B on the gene expressions. Each factor is divided into categories given and not given. Use all 20 genes.

Solution: TwoWayAnovaCh4Ex5.vals has the gene data split into 4 groups: AB, A, B, no input. When we ran TwoWayAnovaSig.py on this input we got:

Observed F-statistic: 22.29
0 out of 10000 experiments had a F-statistic greater than or equal to 22.29
Probability that chance alone gave us a F-statistic of 22.29 or more is 0.0

Note that if you are reporting a p-value of 0 when you do 10,000 tests you should say $p < 0.0001$.

When we ran TwoWayAnovaConf.py on the input file we got:

Observed F-statistic: 22.29
We have 90.0 % confidence that the true F-statistic is between: 9.53 and 41.58
***** Bias Corrected Confidence Interval *****
We have 90.0 % confidence that the true F-statistic is between: 9.24 and 40.85

Remember that the f-statistic measures how the between group variance compares to the within group variance. It looks here that there is between group variance.

6.   Get a Bonferroni corrected p-value using the results from problem 5. Use a Bonferroni cutoff of 0.05.

Solution: The file MultipleTestingCh4Ex6.vals contains the sorted results from problem 3. We multiply each *p*-value by the number of tests run, and then apply the new cutoff. Our new *p*-values are in the last column:

'AB gene 1' and 'A gene 1'	−185.78	0.00000000	0
'AB gene 1' and 'no input gene 1'	−179.49	0.00000000	0
'AB gene 10' and 'no input gene 10'	−182.50	0.00000000	0
'AB gene 11' and 'no input gene 11'	−208.43	0.00000000	0
'AB gene 12' and 'no input gene 12'	−276.83	0.00000000	0
'AB gene 13' and 'A gene 13'	−200.27	0.00000000	0
'AB gene 13' and 'no input gene 13'	−234.84	0.00000000	0
'AB gene 15' and 'no input gene 15'	−204.85	0.00000000	0
'AB gene 16' and 'no input gene 16'	−240.79	0.00000000	0
'AB gene 17' and 'no input gene 17'	−226.53	0.00000000	0
'AB gene 18' and 'A gene 18'	−157.78	0.00000000	0
'AB gene 18' and 'no input gene 18'	−205.78	0.00000000	0
'AB gene 2' and 'no input gene 2'	−192.39	0.00000000	0
'AB gene 3' and 'no input gene 3'	−190.36	0.00000000	0
'AB gene 4' and 'no input gene 4'	−216.10	0.00000000	0
'AB gene 5' and 'A gene 5'	−208.98	0.00000000	0
'AB gene 5' and 'no input gene 5'	−204.71	0.00000000	0
'AB gene 6' and 'no input gene 6'	−208.44	0.00000000	0
'AB gene 7' and 'no input gene 7'	−271.17	0.00000000	0
'AB gene 8' and 'no input gene 8'	−226.06	0.00000000	0
'AB gene 9' and 'no input gene 9'	−165.04	0.00000000	0
'AB gene 9' and 'A gene 9'	43.45	0.01870000	1.122
'AB gene 4' and 'B gene 4'	−43.51	0.01880000	1.128
'AB gene 7' and 'B gene 7'	−36.30	0.03250000	1.95
'AB gene 3' and 'A gene 3'	35.78	0.03500000	2.1
'AB gene 10' and 'A gene 10'	35.63	0.04840000	2.904
'AB gene 20' and 'B gene 20'	56.48	0.07140000	4.284
'AB gene 12' and 'A gene 12'	−28.67	0.08520000	5.112
'AB gene 4' and 'A gene 4'	−24.83	0.10820000	6.492
'AB gene 14' and 'no input gene 14'	−46.64	0.10910000	6.546
'AB gene 6' and 'B gene 6'	−21.75	0.12820000	7.692
'AB gene 12' and 'B gene 12'	−18.81	0.18210000	10.926
'AB gene 20' and 'A gene 20'	32.18	0.19120000	11.472
'AB gene 8' and 'B gene 8'	−18.10	0.19290000	11.574
'AB gene 6' and 'A gene 6'	−17.10	0.19820000	11.892
'AB gene 19' and 'B gene 19'	−30.65	0.21000000	12.6
'AB gene 14' and 'B gene 14'	−27.23	0.22470000	13.482
'AB gene 1' and 'B gene 1'	−15.22	0.22650000	13.59
'AB gene 15' and 'A gene 15'	−14.52	0.24100000	14.46
'AB gene 7' and 'A gene 7'	−12.25	0.27240000	16.344
'AB gene 9' and 'B gene 9'	11.86	0.27870000	16.722
'AB gene 2' and 'A gene 2'	−11.05	0.29720000	17.832
'AB gene 16' and 'B gene 16'	10.80	0.30080000	18.048
'AB gene 17' and 'B gene 17'	−9.58	0.31120000	18.672
'AB gene 19' and 'A gene 19'	17.71	0.32270000	19.362
'AB gene 11' and 'B gene 11'	−8.52	0.34990000	20.994
'AB gene 8' and 'A gene 8'	3.94	0.41670000	25.002
'AB gene 15' and 'B gene 15'	−3.95	0.42260000	25.356
'AB gene 19' and 'no input gene 19'	−7.87	0.42340000	25.404
'AB gene 2' and 'B gene 2'	4.01	0.43110000	25.866

'AB gene 16' and 'A gene 16'	−2.90	0.43840000	26.304
'AB gene 10' and 'B gene 10'	2.85	0.45110000	27.066
'AB gene 5' and 'B gene 5'	2.73	0.45490000	27.294
'AB gene 11' and 'A gene 11'	1.86	0.46870000	28.122
'AB gene 17' and 'A gene 17'	1.19	0.47470000	28.482
'AB gene 3' and 'B gene 3'	1.00	0.48220000	28.932
'AB gene 14' and 'A gene 14'	1.57	0.48520000	29.112
'AB gene 13' and 'B gene 13'	0.95	0.48720000	29.232
'AB gene 18' and 'B gene 18'	−0.80	0.48950000	29.37
'AB gene 20' and 'no input gene 20'	−0.47	0.48990000	29.394

Now when we use a Bonferroni cutoff of 0.05, we reject the null hypothesis in the first 21 tests (we no longer reject the null hypothesis for five of the genes for which we rejected the null hypothesis before). The reason is that when you do so many tests you expect to see some unlikely values by chance.

CHAPTER 5

Case Study: New Mexico's 2004 Presidential Ballots

In our experience, knowing how to approach a problem statistically (understanding the data, figuring out what your question is, and deciding which test should be applied) is much more difficult for people than understanding how to run a pre-specified statistical test when given clean data. In practice, data is rarely as clean as data given in an example used for teaching purposes. Here we take you step by step through a complete data analysis on real data.

New Mexico was one of the key states in the 2004 presidential election and was an important victory for George W. Bush. Because 2004 was a close election and one of the first that made heavy use of electronic voting machines, many people wondered whether one voting machine was clearly better than another. Of course there are many criteria relevant to quality, one being mistakes in the recording of votes. That is the one we discuss here. We do not by contrast discuss the potential for manipulating a vote once inside the voting machine (but others do, see http://accurate-voting.org/pubs/). We analyze the effects of voting machines and ethnicity in those elections.

5.1 Take a close look at the data

First we want to try to understand our data.
[download data]

We must clearly define everything and figure out not just what a term means, but exactly how it was defined during data collection.

Each ballot cast could have been done in one of three ways: early vote, election day vote, and absentee vote. We call these *voting types*.

For holding elections, each county in a state is divided into precincts. Each county can choose the voting technology it uses. Often, a county uses different voting technologies for each of the three voting types specified above.

For each voting type, there are several possible sources of error. An *undervote* is a ballot that does not contain a selection for president. The voter may omit the vote intentionally or the undervote may result from a failure of a voting machine.

A *phantom vote* is a vote for president where there was no actual ballot cast. It can only be a mistake; there should never be more votes for president than ballots cast.

The following is a contrived scenario:

```
100 Ballots
20  undervotes
       11 people opted not to vote for president,
       but did vote in the election
       9 votes for president were lost
5   phantom votes
```

This means that of the 89 people who chose to vote for a presidential candidate, 9 had their votes lost. In addition, 5 votes for president were recorded that shouldn't have been.

Since we don't have the original data, just the summary information that was recorded, the above scenario would show the following result:

```
100 Ballots
85  Votes Cast for President
    (80 successfully recorded real votes + 5 phantom votes = 85 recorded votes for
president)
```

From this we would compute that at least the following irregularities have occurred:

```
15  undervotes (85 - 100 = -15, which is less than 0, and 100 - 85 = 15)
0   phantom votes (85 - 100 = -15, which is not greater than 0, so 0)
```

As you can see, when <u>undervotes</u> and <u>phantom votes</u> are counted together they cancel each other out. Also, we won't know if an undervote was a mistake or was intentional. In order to reduce the canceling out effect, we want to count undervotes and phantom votes separately. That is, instead of taking summary information for the entire state (where either all the phantom votes will be canceled out by the undervotes or vice versa), we count undervotes and phantom votes at the precinct level so that hopefully less canceling out will occur. Even better, since we have summary information for the three different <u>types of votes</u> by precinct, we can use this set of summary information. Ideally, we could get down to the smallest set, the ballot. A ballot can only either have a vote for president, not have a vote for president, or the ballot could be fake, and therefore a phantom vote. If we had this level of information no canceling out would occur (because one ballot cast for president cannot be both an undervote and a phantom vote). As mentioned, we can't know if an individual undervote is a mistake or not. However, according to Liddle et al., undervote rates over 2% usually warrant investigation and so we can keep an eye out for this at the precinct level.

The <u>undervotes</u> for one <u>voting type</u> (early votes) in one precinct (precinct A, an abstract precinct) were counted by:

```
Early Vote undervote in Precinct A =
      max(0, Total Early ballots cast in Precinct A
        - Total Early Presidential Votes cast in Precinct A)

Early Vote phantom votes in Precinct A =
      max(0, Total Early Vote Presidential Votes Cast for Precinct A
        - Total Early Vote Ballots Cast for Precinct A)
```

The above holds for each <u>voting type</u> and each precinct.

```
Total undervotes for Precinct A = Early Vote undervote + Election Day undervote +
Absentee undervote

Total phantom votes = Early Vote phantom vote + Election Day phantom vote +
Absentee undervote
```

If you are still confused about how <u>undervotes</u> and <u>phantom votes</u> can cancel each other out make up a few scenarios yourself and start counting.

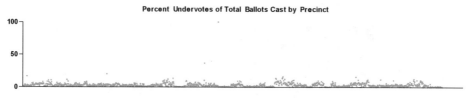

Percent Undervotes of Total Ballots Cast by Precinct

Figure 5.1:   Percent undervotes of total ballots cast by precinct.

The graph below shows the percentage of <u>undervotes</u> for each precinct. These numbers are plotted on a scatter plot above. 1429 precincts reported ballots cast. Notice that according to this graph there is one precinct where 100% of the total ballots were undervotes. This is very suspicious, so we go back to the data to learn more. We find that in precinct 999 of Dona Ana county (a precinct composed of the county's overseas absentee ballots) all 207 of the 207 ballots cast were undervotes. This looks like an outlier, but we do not need to think about whether or not to remove it because we will do a <u>rank transformation</u> on the data. Why we do this is explained <u>later</u>.

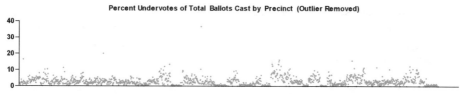

Percent Undervotes of Total Ballots Cast by Precinct (Outlier Removed)

Figure 5.2:   Percent undervotes of total ballots cast by precinct (outlier removed).

Here we have temporarily removed the (possible) outlier in order to see the spread of fraction of <u>undervote</u>s for each precinct more clearly. Notice that the scale of the graph has changed. Glancing at this graph it appears that more than half of the precincts in New Mexico had over 2% undervotes. Remember undervote rates over 2% usually warrant investigation. In fact, 819 precincts had more than 2% undervotes, and 32 precincts had more than 10% undervotes.

Along the x-axis of the graph below are the percentage of undervotes for ballots cast. Displayed are the number of precincts for each range of fraction of undervotes. Here we can see for the majority of the precincts between 1% and 2% of the ballots cast were undervotes. We can see that there are many precincts with no undervotes which means a normal approximation does not work (because, for starters, there is a spike at zero.).

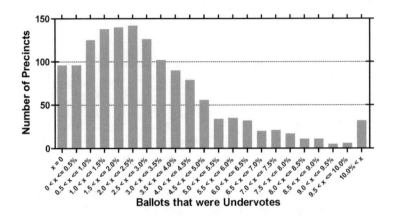

Figure 5.3:   Ballots that were undervotes.

Next we look at the fraction of <u>undervotes</u> by <u>voting type</u>, machine type, and ethnic composition to see if anything interesting comes up.

BE CAREFUL

We shouldn't draw any conclusions without analyzing the data statistically.

In Fig. 5.4, it looks as if there is an association between undervotes and voting type (i.e., it looks as if election day voting is related to an increase in undervotes) as well as between undervotes and machine type in Fig. 5.5 (i.e., it looks as if use of the push button machines, Danaher Shouptronic and Sequoia Advantage, are related to an increase in undervotes).

- Push button DRE - Danaher Shouptronic and Sequoia Advantage
- Touch screen DRE - Sequoia Edge and ES&S iVotronic
- Optical scan (paper ballots) - Optech

Of course, the type of machine used to record votes is not the only thing that distinguishes New Mexico's precincts. Each precinct has a different composition of people as well. The precincts differ in ethnicity, education, income, distance of voter from polling station, etc. Any of these factors could be related to the difference in <u>undervote</u> rates between precincts.

Figure 5.6 shows New Mexico's ethnic composition. Anglos make up about 45%, Hispanics 42%, and Native Americans 9%. Blacks, Asians, and multi-racials, not shown, make up about 2%, 1%, and 1.5% of the population, respectively.

The next two graphs show the ethnic composition for selected precincts. The first shows the 10 precincts that had the smallest undervote rate (but did have at least one undervote each). The second shows the 10 precincts that had the largest undervote rate. In almost all of the precincts which had the smallest <u>undervote</u> rate, Anglos are by far the largest ethnic group and in almost all the precincts which had the largest undervote rate, Hispanics or Native Americans are the largest eth-

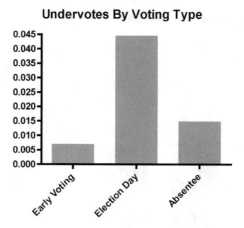

Figure 5.4:   Undervotes by voting type.

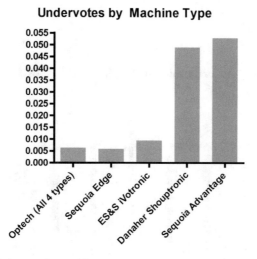

Figure 5.5:   Undervotes by machine type.

nic group. This is worth noting because in a democracy it is important not to disenfranchise part of the population.

See ⊟ for a closer examination of this data..

5.1.1   What questions do we want to ask?

Of the many questions we could ask we are going to try to answer:

Is there a strong association between <u>undervotes</u> and type of voting machine?

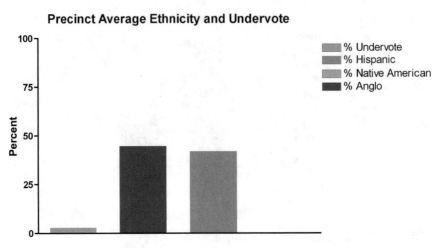

Figure 5.6:   New Mexico's ethnic composition.

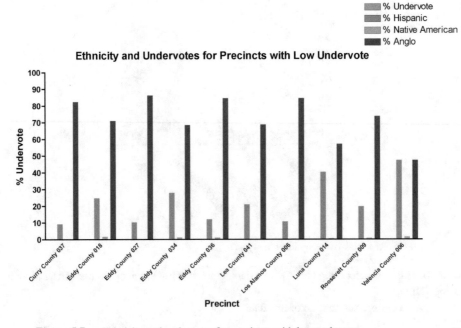

Figure 5.7:   Ethnicity and undervotes for precincts with low undervote.

5.1.2   How do we attempt to answer this question?

To decide whether or not there is a strong association between undervotes and some kind of machine, we will do a one-way ANOVA. We categorize the data on machine type. Therefore the

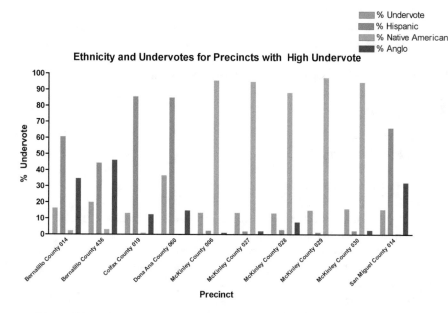

Figure 5.8:   Ethnicity and undervotes for precincts with high undervote.

groups are Push button DRE, Touch screen DRE, and Optical scan. The independent variable is the undervote rank.

We only look at election day data to eliminate <u>voting type</u> as a factor influencing <u>undervote</u> rate. Any precinct with 0 election day ballots is excluded from this test. We look at the fraction of undervotes in each precinct rather than the number of undervotes in each precinct, because 10 undervotes in a precinct where 20 ballots were submitted is very different from 10 undervotes out of 1,000 ballots. We do a rank transformation on the undervote rate to avoid having any outliers significantly distort the result. To assign rankings, the largest undervote rate was ranked 1, the second largest 2, etc. and the undervote rates for each precinct were sorted in descending order (largest first, smallest last). If multiple precincts had the same undervote rate, these precincts were all assigned the same rank. If there are x precincts with undervote rate u and r is the next unused rank, all x precincts are assigned rank $(r + ((r - 1) + x)) / 2$. The next unused rank is then $r + x$. For example, if ranks 1 through 3 have already been assigned, and the next 3 precincts have the same undervote rate, all 3 precincts are ranked $(4 + (3 + 3)) / 2 = 5$, and the next precinct is ranked 7 (assuming it has a unique undervote rate).

Pseudocode & code

All code was tested using Python 2.3.

<u>download code and input files</u>

When we ran the OneWayAnovaSig.py code on the example above we got:

```
Observed F-statistic: 217.29
0 out of 10000 experiments had a F-statistic greater than or equal to 217.29
Probability that chance alone gave us a F-statistic of 217.29 or more is 0.0
```

OK, the difference may be significant, but is it large? Does it matter?

All code was tested using Python 2.3.

<u>download code and input files</u>

When we ran the OneWayAnovaConf.py code we got:

```
Observed F-statistic: 217.29
We have 90.0 % confidence that the true F-statistic is between: 179.85 and 260.16
***** Bias Corrected Confidence Interval *****
We have 90.0 % confidence that the true F-statistic is between: 178.74 and 258.85
```

We conclude that machine type was a significant *predictor* of <u>undervotes</u>. However, this does not necessarily mean that the machine type *caused* the undervote. The undervote could be caused by another factor that happens to be associated with machine type. For example, all precincts in a county use the same machine type on election day and it could be that county demographics are the cause of the difference in undervote numbers.

5.1.3   Next: effect of ethnicity for each machine type

We concluded that machine type is a significant predictor of <u>undervotes</u>, but since machine type is uniform across a county, we next try to remove the effect of county demographics. There are lots of variables related to county that we could examine, but since we have ethnicity information from the 2000 Census that is what we will look at.

We will do three <u>linear regressions</u>, one for each machine type. Each regression will compare ethnicity, in this case we measure the percent of the county that is minority (i.e., non-Anglo), to undervote rate.

First we look at push-button machines.

Pseudocode & code

All code was tested using Python 2.3.

<u>download code and input files</u>

When we ran the RegressionSig.py code we got:

```
Line of best fit for observed data:  y' = 0.0681 x +  1.5328
0 out of 10000 experiments had a slope greater than or equal to 0.0681 .
The chance of getting a slope greater than or equal to 0.0681 is 0.0 .
```

The slope is quite large: an increase in percent minority from the minimum of 0% to the maximum of 100% increases the percent undervote by 6.8%, well above the 2% indication that usually warrants investigation. When ethnicity is strongly correlated with vote selection, then an undervote change of this magnitude can change the outcome of an election. Moreover, the result is significant.

Next we get a confidence interval for the slope.

Pseudocode & code

All code was tested using Python 2.3.

download code and input files

When we ran the RegressionSig.py code we got:

```
Line of best fit for observed data:  y' = 0.0681 x +  1.5328
We have 95.0 % confidence that the true slope is between: 0.0614 and 0.0749
***** Bias Corrected Confidence Interval *****
We have 95.0 % confidence that the true slope is between: 0.0646 and 0.0715
```

Notice that we have 95% confidence that the slope is positive. Next we check how well ethnicity predicts undervote rate.

Pseudocode & code

All code was tested using Python 2.3.

download code and input files

When we ran the CorrelationSig.py code we got:

```
Observed r: 0.52
0 out of 10000 experiments had a r greater than or equal to 0.52
Probability that chance alone gave us a r greater than or equal to 0.52 is 0.00
```

For push-button machines, ethnicity doesn't determine undervote rank, but is a good predictor. Next we do the same for optical machines.

Pseudocode & code

All code was tested using Python 2.3.

download code and input files

When we ran the RegressionSig.py code we got:

```
Line of best fit for observed data:  y' = 0.0249 x +  0.5090
4 out of 10000 experiments had a slope greater than or equal to 0.0249 .
```

```
The chance of getting a slope greater than or equal to 0.0249 is 0.0004 .
```

For optical machines, the slope is very small so a change in ethnicity is only slightly related to a change in undervote rate. However, the result is significant.

Next we get a confidence interval for the slope.

Pseudocode & code

All code was tested using Python 2.3.

<u>download code and input files</u>

When we ran the RegressionConf.py code we got:

```
Line of best fit for observed data:  y' = 0.0249 x +  0.5090
We have 90.0 % confidence that the true slope is between: 0.0137 and 0.0367
***** Bias Corrected Confidence Interval *****
We have 90.0 % confidence that the true slope is between: 0.0138 and 0.0368
```

Again, notice that we have 95% confidence that the slope is positive. Next we check how well ethnicity predicts undervote rate.

Pseudocode & code

All code was tested using Python 2.3.

<u>download code and input files</u>

When we ran the CorrelationSig.py code we got:

```
Observed r: 0.24
2 out of 10000 experiments had a r greater than or equal to 0.24
Probability that chance alone gave us a r greater than or equal to 0.24 is 0.00
```

Ethnicity isn't a great predictor of undervote rank on optical machines.

Finally, we do the same for touch screen machines.

Pseudocode & code

All code was tested using Python 2.3.

<u>download code and input files</u>

When we ran the RegressionSig.py code we got:

```
Line of best fit for observed data:  y' = -0.0659 x +  5.5577
0 out of 10000 experiments had a slope less than or equal to -0.0659 .
The chance of getting a slope less than or equal to -0.0659 is 0.0 .
```

The slope is fairly large and is negative this time. Also, the result is statistically significant. Next we get a confidence interval for the slope.

Pseudocode & code

All code was tested using Python 2.3.

<u>download code and input files</u>

When we ran the RegressionConf.py code we got:

```
Line of best fit for observed data:  y' = -0.0659 x +  5.5577
We have 95.0 % confidence that the true slope is between: -0.0927 and -0.0421
***** Bias Corrected Confidence Interval *****
We have 95.0 % confidence that the true slope is between: -0.0790 and -0.0540
```

Notice that here we have 95% confidence that the slope is *negative*. Next we check how well ethnicity predicts undervote rate.

Pseudocode & code

All code was tested using Python 2.3.

<u>download code and input files</u>

When we ran the CorrelationSig.py code we got:

```
Observed r: -0.52
0 out of 10000 experiments had a r less than or equal to -0.52
Probability that chance alone gave us a r less than or equal to -0.52 is 0.00
```
Ethnicity isn't a great predictor of undervote rank, but it isn't bad.

Now let's look at the results of this analysis in a graph:

If the slopes of these lines had all been uniformly negative or positive, we would have suspected that ethnicity had a strong effect on undervote counts. This does not hold. So, ethnicity is no proxy for tendency to undervote. The next question is whether the voting machine itself could determine the tendency to undervote. To minimize the effect of ethnicity, we can split the data into quadrants based on percent minority, and then for each quadrant ask if the machine types are significantly different. The purpose of doing this would be to look at sections of the data in which percent minorty does not vary much. At the end of this exercise we may be able to say, in the case where percent minority is between *a* and *b*, push button, touch screen, and optical scan machines do/do not signif-icanlty differ.

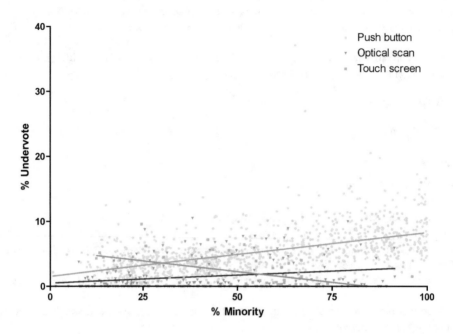

Figure 5.9:   Percent minority comparison between push button, touch screen, and optical scan machines.

The next question is whether the voting machine itself could determine the tendency to undervote. To minimize the effect of ethnicity, we can split the data into quadrants based on percent minority, and then for each quadrant ask if the machine types are significantly different. The purpose of doing this would be to look at sections of the data in which percent minority does not vary much. At the end of this exercise we may be able to say, in the case where percent minority is between a and b, push button, touch screen, and optical scan machines do/do not significantly differ.

Since we want to determine if the three machine types significantly differ in each of these four cases, we will do four one-way ANOVAs. First we look at precincts that have a percent minority greater than or equal to zero and less than 25.

Pseudocode & code

All code was tested using Python 2.3.

download code and input files

When we ran the OneWayAnovaSig.py code we got:

```
Observed F-statistic: 61.12
0 out of 10000 experiments had a F-statistic greater than or equal to 61.12
Probability that chance alone gave us a F-statistic of 61.12 or more is 0.0
```

Pseudocode & code

All code was tested using Python 2.3.

download code and input files

When we ran the OneWayAnovaConf.py code we got:

```
Observed F-statistic: 61.12
We have 90.0 % confidence that the true F-statistic is between: 47.68 and 79.64
***** Bias Corrected Confidence Interval *****
We have 90.0 % confidence that the true F-statistic is between: 45.33 and 76.74
```

Now we look at the precincts that have percent minority greater than or equal to 25 and less than 50.

Pseudocode & code

All code was tested using Python 2.3.

download code and input files

When we ran the OneWayAnovaSig.py code we got:

```
Observed F-statistic: 36.21
0 out of 10000 experiments had a F-statistic greater than or equal to 36.21
Probability that chance alone gave us a F-statistic of 36.21 or more is 0.0
```

Pseudocode & code

All code was tested using Python 2.3.

download code and input files

When we ran the OneWayAnovaConf.py code we got:

```
Observed F-statistic: 36.21
We have 90.0 % confidence that the true F-statistic is between: 21.70 and 57.96
***** Bias Corrected Confidence Interval *****
We have 90.0 % confidence that the true F-statistic is between: 19.83 and 55.18
```

Next we look at precincts where the percent minority is greater than or equal to 50 and less than 75.

Pseudocode & code

All code was tested using Python 2.3.

download code and input files

When we ran the OneWayAnovaSig.py code we got:

```
Observed F-statistic: 61.70
0 out of 10000 experiments had a F-statistic greater than or equal to 61.70
Probability that chance alone gave us a F-statistic of 61.70 or more is 0.0
```

Pseudocode & code

All code was tested using Python 2.3.

download code and input files

When we ran the OneWayAnovaConf.py code we got:

```
Observed F-statistic: 61.70
We have 90.0 % confidence that the true F-statistic is between: 43.77 and 87.03
***** Bias Corrected Confidence Interval *****
We have 90.0 % confidence that the true F-statistic is between: 42.12 and 83.99
```

Finally, we look at precincts where the percent minority is greater than or equal to 75 and less than 100.

Pseudocode & code

All code was tested using Python 2.3.

download code and input files

When we ran the OneWayAnovaSig.py code we got:

```
Observed F-statistic: 74.62
0 out of 10000 experiments had a F-statistic greater than or equal to 74.62
Probability that chance alone gave us a F-statistic of 74.62 or more is 0.0
```

Pseudocode & code

All code was tested using Python 2.3.

<u>download code and input files</u>

When we ran the OneWayAnovaConf.py code we got:

```
Observed F-statistic: 74.62
We have 90.0 % confidence that the true F-statistic is between: 45.87 and 116.16
***** Bias Corrected Confidence Interval *****
We have 90.0 % confidence that the true F-statistic is between: 43.64 and 112.41
```

We conclude that in all four quadrants the undervote rates differ significantly for different machine types. Note that this does not mean that in any given quadrant machine type *a* is significantly different from machine type *b*, only that the three machines types differ. To determine if any two machine types are significantly different in a quadrant, we must compare those two machine types. In addition, we have not said which machine type performs the best in each quadrant. To do that we would have to compare the means.

Remember we have not shown that a change in percent minority *causes* the change in the expected undervote rate for a particular machine, we are claiming only that a change in percent minority is correlated with a change in the undervote rate for a particular machine. It is an important distinction to make. Any number of things could be causing the change in undervote rate for a machine type: income, distance from polling station, quality of public education, comfort level with computers, etc.

5.1.4   We have used the following techniques:

To measure the difference in undervotes for the three machine types we used a <u>One-way ANOVA</u>. We did a rank transformation on the undervote rate to avoid having outliers distort the result of the one-way ANOVA.

Since each county in New Mexico used one machine type on election day, instead of the machine types being randomly distributed across the state, we attempted to remove the effect of a county demographic, ethnicity, to compare machine type and undervote rate. To get an idea of how ethnicity is related to the relationship between undervotes and machine types, we ran three <u>linear regressions</u>, one for each machine type. We then overlaid the three regression lines in a graph to compare them.

The effects of machine type and ethnicity are interrelated. To see if the three machine types significantly differed in subsections of the population where the variance in percent minority is smaller, we split the precincts up into four quadrants based on percent minority and then ran four <u>One-way ANOVAs</u>.

5.1.5   What did we find out?

One-way ANOVA suggests that certain machine types lead to more undervotes, but which machine type performs the best depends on the demographics of the area. In fact, the voters using touch screen machines suffered the least undervotes when percent non-Anglo was high (over approximately 55% of the population), but, in all other cases, voters using optical scan machines suffered the least undervotes. So, it is not true that "a single machine type is consistently more likely to yield undervotes over all ethnicity levels."

There would still be other questions to ask such as the degree of repair of the voting machines in the different countries, the model, the number of machines, the length of waiting lines. We do not have that information, but if you do, then you can use a similar analysis.

References

Andrews, D. F. F., Bickel, P. J., Hampel, F. R., Tukey, J. W., and Huber, P. J., *Robust Estimates of Location: Survey and Advances*, Princeton University Press, Princeton, 1972.

Bruce, P. C. and Simon, J. L., "The Statistical Results of Resampling Instruction: Evaluations of Teaching Introductory Statistics via Resampling," *Resampling Stats*. <http://www.resample.com/content/teaching/texts/i-1reslt.txt>

Campbell, M. K. and Torgerson, D. J., "Bootstrapping: estimating confidence intervals for cost-effectiveness ratios," *Quarterly Journal of Medicine,* 92:177-182, 1999. <http://qjmed.oxford-journals.org/cgi/content/full/92/3/177>

Chiaromonte, F., "Statistical Analysis of Genomics Data," New York University Courant Institute of Mathematical Sciences, New York, <http://www.cims.nyu.edu/~chiaro/SAGD_05>

(An excellent introduction to resampling in the context of biological problems.)

Davison, A.C. and Hinkley, D. V., *Bootstrap Methods and their Application*. Cambridge University Press, Cambridge, 1997.

"Distribution Tables," StatSoft. <http://www.statsoft.com/textbook/sttable.html>

Easton, V. J. and McColl, J. H., "Design of Experiments and ANOVA," *Statistics Glossary*. <http://www.cas.lancs.ac.uk/glossary_v1.1/dexanova.html>

Good, P. I., *Resampling Methods: A Practical Guide to Data Analysis*, Birkhäuser, Boston, 2006.

Howell, D., "Resampling Statistics: Randomization and the Bootstrap." The University of Vermont, Burlington, VT. <http://www.uvm.edu/~dhowell/StatPages/Resampling/Resampling.html>

Liddle, E., Mitteldorf, J., Sheehan, R. G., and Baiman, R., "Analysis of Undervotes in New Mexico's 2004 Presidential Ballots," National Election Data Archive. Jan. 2005. <http://uscountvotes.org/ucvAnalysis/NM/NMAnalysis_EL_JM.pdf>

Lowry, R., *Concepts and Applications of Inferential Statistics*. <http://faculty.vassar.edu/lowry/webtext.html>

Lunneborg, C. E., *Data Analysis by Resampling: Concepts and Applications*, Duxbury, 2000.

Lunneborg, C. E., "Modeling Experimental and Observational Data", iUniverse.com, 2000.

"Muriel Bristol." Wikipedia. <http://en.wikipedia.org/wiki/Muriel_Bristol>

"Outlier." Wikipedia. <http://en.wikipedia.org/wiki/Outlier>

Sen, S., "Multiple Comparisons and the False Discovery Rate." University of California San Francisco. <http://www.biostat.ucsf.edu/biostat/sen/papers/fdr.pdf>

Taleb, N., "Learning to Expect the Unexpected." *Edge*. 14 Apr. 2004. <http://www.edge.org/3rd_culture/taleb04/taleb_index.html>

Theisen, E. and Stewart, W., "Summary Report on New Mexico State Election Data." Democracy for New Mexico. www.HelpAmericaRecount.Org. January 4, 2005. <http://www.democracy-fornewmexico.com/democracy_for_new_mexico/files/NewMexico2004ElectionDataReport-v2.pdf>

Walters, S. J., "Sample size and power estimation for studies with health related quality of life outcomes: a comparison of four methods using the SF-36," *Health Qual Life Outcomes*, 2:26, 2004.. <http://www.pubmedcentral.nih.gov/articlerender.fcgi?artid=421748>

Weaver, B., Statistics Notes. <http://www.angelfire.com/wv/bwhomedir/notes.html>

CHAPTER A

Bias Corrected Confidence Intervals

Our method of computing <u>confidence intervals</u> works well when our original sample is the median of the bootstrap values. When this is not the case, we say the data is "biased." Bias often arises because extreme values in the data are on one side only, e.g., a collection of wind speed values including a few from hurricanes. We can correct for bias by computing how far the original sample value is from the mean of our bootstrap values, and adjusting our interval accordingly. Resampling pioneer Bradley Efron calls this a "bias corrected confidence interval" ⊟.

First, we find the fraction of <u>bootstraps</u> below the value computed from the original sample (that value is hereafter called the *original sample estimate*). To do this, order all the bootstrap sample estimates, count the number of them that are below the original sample estimate, then divide that by the total number of bootstrap estimates. So, if we did 10,000 bootstraps and 4,500 of them have a sample estimate below our original sample estimate, then 4,500/10,000 = 0.45 (45% of the bootstrap estimates are below our whole estimate). Intuitively, this means that the estimate based on the whole sample is less than the median of the bootstrapping. So, we'll have to adjust the confidence interval downwards because the estimate based on the whole sample should get more weight than the bootstraps. Intuition has told us in which direction to adjust the confidence interval. A slightly involved statistical algorithm will now tell us how much. The python code associated with each statistic discussed later implements this algorithm as one option.

Imagine a normal curve. The mean is the same as the median that is greater than 50% of the values. A point that is greater than only 45% of the values is shown just to the left. A fraction 0.05 of all values are in between (see Fig. 2.1). We can find out how many standard deviations away the 0.45 point is by consulting the ϕ table below.

The left column of the table represents fractions of standard deviations in increments of 0.1. The column headers represent additional fractions of standard deviations in increments of 0.01. Thus, the cell where the row 0.1 and the column 0.02 intersect corresponds to 0.12 standard deviations. In that cell, we find 0.0478. This expresses the function ϕ (0.12) = 0.0478. So, under a normal

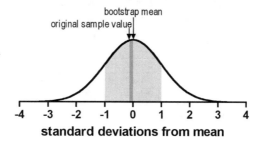

Figure A.1: Standard deviations from mean.

assumption, a value that is greater than $0.5 + 0.0478$ ($= 0.5478$) of the values is 0.12 standard deviations above the mean. A value that is greater than $0.5 + 0.0517$ ($= 0.5517$) is 0.13 standard deviations above the mean. Interpolating, we see that a value that is greater than 0.55 of the values is approximately 0.125 standard deviations above the mean. In our case, we are talking about a value that is greater than 0.45 of the values (that is, a value 0.05 *below* the mean) so its value in standard deviations is symmetrically negative, i.e., -0.125. We call that value the bias correcting constant z_0. In summary, the table computes the function ϕ from standard deviations (in bold) to fractions of values (in plaintext in the cells) above the median. It is symmetric for values below the median.

Area between the mean and a value x standard deviations above the mean[*]

	0.00	0.01	0.02	0.03	0.04	0.05	0.06	0.07	0.08	0.09
0.0	0.0000	0.0040	0.0080	0.0120	0.0160	0.0199	0.0239	0.0279	0.0319	0.0359
0.1	0.0398	0.0438	**0.0478**	**0.0517**	0.0557	0.0596	0.0636	0.0675	0.0714	0.0753
0.2	0.0793	0.0832	0.0871	0.0910	0.0948	0.0987	0.1026	0.1064	0.1103	0.1141
0.3	0.1179	0.1217	0.1255	0.1293	0.1331	0.1368	0.1406	0.1443	0.1480	0.1517
0.4	0.1554	0.1591	0.1628	0.1664	0.1700	0.1736	0.1772	0.1808	0.1844	0.1879
0.5	0.1915	0.1950	0.1985	0.2019	0.2054	0.2088	0.2123	0.2157	0.2190	0.2224
0.6	0.2257	0.2291	0.2324	0.2357	0.2389	0.2422	0.2454	0.2486	0.2517	0.2549
0.7	0.2580	0.2611	0.2642	0.2673	0.2704	0.2734	0.2764	0.2794	0.2823	0.2852
0.8	0.2881	0.2910	0.2939	0.2967	0.2995	0.3023	0.3051	0.3078	0.3106	0.3133
0.9	0.3159	0.3186	0.3212	0.3238	0.3264	0.3289	0.3315	0.3340	0.3365	0.3389
1.0	0.3413	0.3438	0.3461	0.3485	0.3508	0.3531	0.3554	0.3577	0.3599	0.3621
1.1	0.3643	0.3665	0.3686	0.3708	0.3729	0.3749	0.3770	0.3790	0.3810	0.3830
1.2	0.3849	0.3869	0.3888	0.3907	0.3925	0.3944	0.3962	0.3980	0.3997	0.4015
1.3	0.4032	0.4049	0.4066	0.4082	0.4099	0.4115	0.4131	0.4147	0.4162	0.4177
1.4	0.4192	0.4207	0.4222	0.4236	0.4251	0.4265	0.4279	0.4292	0.4306	0.4319
1.5	0.4332	0.4345	0.4357	0.4370	0.4382	0.4394	0.4406	0.4418	0.4429	0.4441
1.6	0.4452	0.4463	0.4474	0.4484	0.4495	0.4505	0.4515	0.4525	0.4535	0.4545
1.7	0.4554	**0.4564**	0.4573	0.4582	0.4591	0.4599	0.4608	0.4616	0.4625	0.4633
1.8	0.4641	0.4649	0.4656	0.4664	0.4671	0.4678	0.4686	0.4693	0.4699	0.4706
1.9	0.4713	0.4719	0.4726	0.4732	0.4738	0.4744	**0.4750**	0.4756	0.4761	0.4767
2.0	0.4772	0.4778	0.4783	0.4788	0.4793	0.4798	0.4803	0.4808	0.4812	0.4817
2.1	0.4821	0.4826	0.4830	0.4834	0.4838	0.4842	0.4846	0.4850	0.4854	0.4857
2.2	0.4861	**0.4864**	0.4868	0.4871	0.4875	0.4878	0.4881	0.4884	0.4887	0.4890
2.3	0.4893	0.4896	0.4898	0.4901	0.4904	0.4906	0.4909	0.4911	0.4913	0.4916
2.4	0.4918	0.4920	0.4922	0.4925	0.4927	0.4929	0.4931	0.4932	0.4934	0.4936
2.5	0.4938	0.4940	0.4941	0.4943	0.4945	0.4946	0.4948	0.4949	0.4951	0.4952
2.6	0.4953	0.4955	0.4956	0.4957	0.4959	0.4960	0.4961	0.4962	0.4963	0.4964
2.7	0.4965	0.4966	0.4967	0.4968	0.4969	0.4970	0.4971	0.4972	0.4973	0.4974
2.8	0.4974	0.4975	0.4976	0.4977	0.4977	0.4978	0.4979	0.4979	0.4980	0.4981
2.9	0.4981	0.4982	0.4982	0.4983	0.4984	0.4984	0.4985	0.4985	0.4986	0.4986
3.0	0.4987	0.4987	0.4987	0.4988	0.4988	0.4989	0.4989	0.4989	0.4990	0.4990

[*]The normal distribution table comes from 🗗.

If you now want a 95% confidence interval, the fraction of bootstraps that we are omitting is 0.05, so the fraction of bootstraps that we are omitting on each end is half that or .025, and we want to go from 2.5% to 97.5%. So, 47.5% of the values should be directly below the mean, and 47.5%

above (because 50% - 2.5% = 47.5%). By looking at the table we find that .475 of the values lie 1.96 standard deviations below the mean, and .475 of the values lie 1.96 standard deviations above the mean. $Z_{\alpha/2} = -1.96$; $Z_{1-\alpha/2} = 1.96$. This is the standard normal deviate associated with $\alpha/2$.

The lower limit of the bias-corrected confidence interval is the value of the bootstrap estimate at the $0.5 + \phi(Z_{\alpha/2} + 2z_0)^{\text{th}}$ percentile and the upper limit is the $0.5 + \phi(Z_{1-\alpha/2} + 2z_0)^{\text{th}}$ percentile. From our example above we have $Z_{\alpha/2} = -1.96$ and $Z_{1-\alpha/2} = 1.96$. Remember that z_0 is the standard normal deviate corresponding to the proportion of bootstrap values below our original sample value. So in our example, $z_0 = -0.125$, $\phi(-1.96 + 2(-0.125)) = \phi(-2.21) = -0.4864$, $\phi(1.96 + 2(-0.125)) = \phi(1.71) = 0.4564$, and so we look at the bootstrap estimates at the $(0.5 + -0.4864)^{\text{th}} = 0.0136^{\text{th}}$ percentile position, and the $(0.5 + 0.4564)^{\text{th}} = 0.9546^{\text{th}}$ percentile position. Since we took 10,000 bootstraps, we have a 95% confidence interval from the bootstrap value at position 136 (we got 136 from $0.0136 \cdot 10,000$) to the bootstrap value at position 9564 (we got 9564 from $0.9564 \cdot 10,000$).

All code was tested using Python 2.3.

<u>download code and input files</u>

Here is an example run of this code, using the Diff2MeanConfCorr.py from our <u>previous</u> placebo versus drug example:

```
Observed difference between the means: 12.97
We have 90.0 % confidence that the true difference between the means is between:
7.60 and 18.12
```

The reason it should be $+2\,z_0$ can be explained intuitively as follows. Imagine a town in which there are relatively few very rich people and the rest are of modest means. Suppose the statistic is the mean. In this town the mean will be greater than the median. In that case the proportion p that are below the sample estimate will be > 0.5. So z_0 will be positive reflecting the fact that the mean is so many standard deviations above the median. (The percentile method would give the same result if the median and mean were the same value so z_0 were 0.) We use this in the calculation to give a little extra weight to those few rich people, because the original sample had them.

Appendix B

All code was tested using Python 2.3.

Coinsig.py

```
#!/usr/bin/python

######################################
# Coin Toss Significance Test
# From: Statistics is Easy! By Dennis Shasha and Manda Wilson
#
#
# Simulates 10,000 experiments, where each experiment is 17 tosses of a fair coin.
# For each experiment we count the number of heads that occur out of the 17 tosses.
# We then compare our observed number of heads in 17 tosses of a real coin to this
# distribution, to determine the probability that our sample came from this assumed
# distribution (to decide whether this coin is fair).
#
# Author: Manda Wilson
#
# Pseudocode:
#
# 1. Set a counter to 0, this will count the number of times we get 15 or more
#    heads out of 17 tosses.
#
# 2. Do the following 10,000 times:
#    a. Create a drawspace (a big pool of numbers we will pick from).
#       In our example, p, the probability of "success", i.e., tossing a fair coin and
#       getting a head, equals .5, and our drawspace will be from 0 to 2000.05.  We get

#       2000.05 from ((1 / 0.5) * 1000) + 0.05.
#    b. Do the following 17 times, counting successes as we go:
#       i. Pick randomly from the drawspace.
#       ii. If the number we drew is less or equal to p * drawspace then this was a
success.
#             p * drawspace is the proportion of values in the drawspace that count as
successes.
#             In our example, p * drawspace equals 1000.025, so if we drew any number
less than or equal to this
#             it counts as a success (success = we drew a head).
```

```
#      c. If the number of successes from step (2b) is greater than or equal to 15,
increment our counter
#          from step (1).
#
# 3. counter / 10,000 equals the probability of getting an observed number of heads
greater than or equal to 15
#     in 17 tosses if the coin is fair.
#
####################################

import random

####################################
#
# Adjustable variables
#
####################################

observed_number_of_heads = 15
number_of_tosses = 17
probability_of_head = 0.5


####################################
#
# Subroutines
#
####################################

# p = probability of some outcome for a trial
# n = number of trials
# returns number of times outcome of probability p occurred out of n trials
def applyprob(p, n):
  drawspace = ((1 / p) * 1000) + 0.05
  success = 0
  for j in range(n):
    outcome = random.uniform(0, drawspace)
    if ( (p * drawspace) >= outcome):
      success = success + 1
  return success

####################################
#
# Computations
#
####################################

countgood = 0
number_of_bootstraps = 10000
out=[]

for i in range(number_of_bootstraps):
  out.append(applyprob(probability_of_head,number_of_tosses))
```

```
# count the number of times we got greater than or equal to 15 heads out of 17 coin
tosses
countgood = len(filter(lambda x: x >= observed_number_of_heads, out))


####################################
#
# Output
#
####################################

print countgood, "out of", number_of_bootstraps, "times we got at least",
print observed_number_of_heads, "heads in", number_of_tosses, "tosses."
print "Probability that chance alone gave us at least", observed_number_of_heads,
print "heads in", number_of_tosses, "tosses is", countgood /
float(number_of_bootstraps), "."
```

Diff2MeanSig.py

```
#!/usr/bin/python


####################################
# Difference between Two Means Significance Test
# From: Statistics is Easy! By Dennis Shasha and Manda Wilson
#
#
# Assuming that there is no significant difference in the means of
# the two samples, tests to see the probability of getting a difference
# greater than or equal to the observed difference in the means by chance
# alone.  Uses shuffling & bootstrapping to get a distribution to compare
# to the observed statistic.
#
# Author: Manda Wilson
#
# Example of FASTA formatted input file (this is only for 2 groups):
# >placebo_vals
# 54 51 58 44 55 52 42 47 58 46
# >drug_vals
# 54 73 53 70 73 68 52 65 65
#
# Pseudocode:
#
# 1. Measure the difference between the two group means.  The difference in means is
measured
#    by (sum(grpA) / len(grpA)) - (sum(grpB) / len(grpB)).  In this example the
difference between
#    the two group means is 12.97.
#
# 2. Set a counter to 0, this will count the number of times we get a difference
#    between the means greater than or equal to 12.97.
#
# 3. Do the following 10,000 times:
#    a. Shuffle the original measurements.  To do this:
```

```
#        i. put the values from all the groups into one array but remembering the start
and end indexes of each group
#        ii. shuffle the values in the array, effectively reassigning the values to
different groups
#     b. Measure the difference between the two group means, just as we did in step (1).
#     c. If the difference from step (3b) is greater than or equal to 12.97, increment
our counter
#        from step (2). Note: if our original difference between the means were a
negative value
#        we would check for values less than or equal to that value.
#
# 4. counter / 10,000 equals the probability of getting our observed difference of two
means greater than
#     or equal to 12.97, if there is in fact no significant difference.
#
####################################

import random

####################################
#
# Adjustable variables
#
####################################

input_file = "input/Diff2Mean.vals"

####################################
#
# Subroutines
#
####################################

# takes a list of groups (two or more)
# pools all values, shuffles them, and makes new groups
# of same size as original groups
# returns these new groups
# example of shuffle with more than two groups: http://www.statisticsiseasy.com/code/
OneWayAnovaSig.py
def shuffle(grps):
    num_grps = len(grps)
    pool = []

    # pool all values
    for i in range(num_grps):
        pool.extend(grps[i])
    # mix them up
    random.shuffle(pool)
    # reassign to groups of same size as original groups
    new_grps = []
    start_index = 0
    end_index = 0
```

```
    for i in range(num_grps):
       end_index = start_index + len(grps[i])
       new_grps.append(pool[start_index:end_index])
       start_index = end_index
    return new_grps

# subtracts group a mean from group b mean and returns result
def meandiff(grpA, grpB):
    return sum(grpB) / float(len(grpB)) - sum(grpA) / float(len(grpA))


####################################
#
# Computations
#
####################################

# list of lists
samples = []
a = 0
b = 1

# file must be in FASTA format
infile=open(input_file)
for line in infile:
    if line.startswith('>'):
       # start of new sample
       samples.append([])
    elif not line.isspace():
       # line must contain values for previous sample
       samples[len(samples) - 1] += map(float,line.split())
infile.close()

observed_mean_diff = meandiff(samples[a], samples[b])


count = 0
num_shuffles = 10000

for i in range(num_shuffles):
    new_samples = shuffle(samples)
    mean_diff = meandiff(new_samples[a], new_samples[b])
    # if the observed difference is negative, look for differences that are smaller
    # if the observed difference is positive, look for differences that are greater
    if observed_mean_diff < 0 and mean_diff <= observed_mean_diff:
       count = count + 1
    elif observed_mean_diff >= 0 and mean_diff >= observed_mean_diff:
       count = count + 1


####################################
#
# Output
#
####################################
```

```
print "Observed difference of two means: %.2f" % observed_mean_diff
print count, "out of", num_shuffles, "experiments had a difference of two means",
if observed_mean_diff < 0:
   print "less than or equal to",
else:
   print "greater than or equal to",
print "%.2f" % observed_mean_diff, "."
print "The chance of getting a difference of two means",
if observed_mean_diff < 0:
   print "less than or equal to",
else:
   print "greater than or equal to",
print "%.2f" % observed_mean_diff, "is", (count / float(num_shuffles)), "."
```

Diff2Mean.vals

```
>placebo_vals
54 51 58 44 55 52 42 47 58 46
>drug_vals
54 73 53 70 73 68 52 65 65
```

Diff2MeanConf.py

```
#!/usr/bin/python

#####################################
# Difference between Two Means Confidence Interval
# From: Statistics is Easy! By Dennis Shasha and Manda Wilson
#
#
# Uses shuffling & bootstrapping to get a 90% confidence interval for the difference
between two means.
#
# Author: Manda Wilson
#
# Example of FASTA formatted input file (this is only for 2 groups):
# >placebo_vals
# 54 51 58 44 55 52 42 47 58 46
# >drug_vals
# 54 73 53 70 73 68 52 65 65
#
# Pseudocode:
#
# 1. Measure the difference between the two group means.  The difference in means is
measured
#    by sum(grpA) / len(grpA) - sum(grpB) / len(grpB).  In this example the difference
between
#    the two group means is 12.97.
#
# 2. Do the following 10,000 times:
#    a. For each sample we have get a bootstrap sample:
#        i. Create a new array of the same size as the original sample
```

```
#       ii. Fill the array with randomly picked values from the original sample
(randomly picked with replacement)
#     b. Measure the difference between the two bootstrap group means, just as we did
in step (1)
#       with the original samples.
#
# 3. Sort the differences computed in step (2).
#
# 4. Compute the size of each interval tail.  If we want a 90% confidence interval,
then 1 - 0.9 yields the
#    portion of the interval in the tails.  We divide this by 2 to get the size of each
tail, in this case 0.05.
#
# 5. Compute the upper and lower bounds.  To get the lower bound we multiply the tail
size by the number of
#    bootstraps we ran and round this value up to the nearest integer (to make sure
this value maps
#    to an actual boostrap we ran).  In this case we multiple 0.05 by 10,000 and get
500, which rounds up to 500.
#
#    To compute the upper bound we subtract the tail size from 1 and then multiply that
by the number of bootstraps
#    we ran.  Round the result of the last step down to the nearest integer.  Again,
this is to ensure this value
#    maps to an actual bootstrap we ran.  We round the lower bound up and the upper
bound down to reduce the
#    confidence interval size, so that we can still say we have as much confidence in
this result.
#
# 6. The bootstrap values at the lower bound and upper bound give us our confidence
interval.
#
####################################

import random
import math
import sys


####################################
#
# Adjustable variables
#
####################################

input_file = "input/Diff2Mean.vals"
conf_interval = 0.9


####################################
#
# Subroutines
#
####################################
```

```
# x is an array of sample values
# returns a new array with randomly picked
# (with replacement) values from x
def bootstrap(x):
   samp_x = []
   for i in range(len(x)):
     samp_x.append(random.choice(x))
   return samp_x

# subtracts group a mean from group b mean and returns result
def meandiff(grpA, grpB):
   return sum(grpB) / float(len(grpB)) - sum(grpA) / float(len(grpA))


####################################
#
# Computations
#
####################################

# list of lists
samples = []
a = 0
b = 1

# file must be in FASTA format
infile=open(input_file)
for line in infile:
   if line.startswith('>'):
     # start of new sample
     samples.append([])
   elif not line.isspace():
     # line must contain values for previous sample
     samples[len(samples) - 1] += map(float,line.split())
infile.close()

observed_mean_diff = meandiff(samples[a], samples[b])

num_resamples = 10000    # number of times we will resample from our original samples
out = []                 # will store results of each time we resample

for i in range(num_resamples):
   # get bootstrap samples for each of our groups
   # then compute our statistic of interest
   # append statistic to out
   bootstrap_samples = []  # list of lists
   for sample in samples:
     bootstrap_samples.append(bootstrap(sample))
   # now we have a list of bootstrap samples, run meandiff
   out.append(meandiff(bootstrap_samples[a], bootstrap_samples[b]))

out.sort()
```

```
tails = (1 - conf_interval) / 2

# in case our lower and upper bounds are not integers,
# we decrease the range (the values we include in our interval),
# so that we can keep the same level of confidence
lower_bound = int(math.ceil(num_resamples * tails))
upper_bound = int(math.floor(num_resamples * (1 - tails)))


###################################
#
# Output
#
###################################


# print observed value and then confidence interval
print "Observed difference between the means: %.2f" % observed_mean_diff
print "We have", conf_interval * 100, "% confidence that the true difference between
the means",
print "is between: %.2f" % out[lower_bound], "and %.2f" % out[upper_bound]
```

Diff2MeanConfCorr.py

```
#!/usr/bin/python


###################################
# Difference between Two Means Bias Corrected Confidence Interval
# From: Statistics is Easy! By Dennis Shasha and Manda Wilson
#
#
# Uses shuffling & bootstrapping to get a 90% confidence interval for the difference
between two means.
# The confidence interval is computed using Efron's bias corrected method.
#
# Author: Manda Wilson
#
# Example of FASTA formatted input file (this is only for 2 groups):
# >placebo_vals
# 54 51 58 44 55 52 42 47 58 46
# >drug_vals
# 54 73 53 70 73 68 52 65 65
#
# Pseudocode:
#
# 1. Measure the difference between the two group means.  The difference in means is
measured
#    by sum(grpA) / len(grpA) - sum(grpB) / len(grpB).  In this example the difference
between
#    the two group means is 12.97.
#
# 2. Do the following 10,000 times:
#    a. For each sample we take a bootstrap sample:
#       i. Create a new array of the same size as the original sample
```

```
#        ii. Fill the array with randomly picked values from the original sample
(uniformly and randomly picked with replacement)
#     b. Measure the difference between the two bootstrap group means, just as we did
in step (1)
#        with the original samples.
#
# 3. Sort the differences computed in step (2).
#
# 4. Compute the size of each interval tail.  If we want a 90% confidence interval,
then 1 - 0.9 yields the
#    portion of the interval in the tails.  We divide this by 2 to get the size of each
tail, in this case 0.05.
#
# 5. Compute the upper and lower bounds.
#     a. Find proportion p of the bootstrap values computed in step (2) that are below
the sample estimate
#        z_0 = phi^-1(p)
#           For example if p = 0.5 then z_0 = 0, if p = 0.5 - 0.3413 then z_0 = -1.
#     b. z_alpha/2 = phi^-1(alpha/2). e.g. phi^-1(0.05) when getting the 90% confidence
interval (i.e. alpha = 10%)
#     c. z_1-alpha/2 = phi^-1(1-alpha/2) e.g. phi^-1(0.95) when getting the 90%
confidence interval
#     d. the lower end of the alpha confidence interval is phi(z_alpha/2 + 2 z_0) among
the bootstrap values
#     e. the upper end is phi(z_1-alpha/2 + 2 z_0)
#
# 6. The bootstrap values at the lower bound and upper bound give us our confidence
interval.
#
#################################

import random
import math
import sys


#################################
#
# Adjustable variables
#
#################################

input_file = "input/Diff2Mean.vals"
conf_interval = 0.9


#################################
#
# Subroutines
#
#################################

# maps proportion of values above mean
# to number of standard deviations above mean
# keys will be index / 100 \:[0-9]\.[0-9][0-9]\,
```

```
area_to_sd_map = [0.0000, 0.0040, 0.0080, 0.0120, 0.0160, 0.0199, 0.0239, 0.0279,
0.0319, 0.0359, 0.0398, 0.0438, 0.0478, 0.0517, 0.0557, 0.0596, 0.0636, 0.0675,
0.0714, 0.0753, 0.0793, 0.0832, 0.0871, 0.0910, 0.0948, 0.0987, 0.1026, 0.1064,
0.1103, 0.1141, 0.1179, 0.1217, 0.1255, 0.1293, 0.1331, 0.1368, 0.1406, 0.1443,
0.1480, 0.1517, 0.1554, 0.1591, 0.1628, 0.1664, 0.1700, 0.1736, 0.1772, 0.1808,
0.1844, 0.1879, 0.1915, 0.1950, 0.1985, 0.2019, 0.2054, 0.2088, 0.2123, 0.2157,
0.2190, 0.2224, 0.2257, 0.2291, 0.2324, 0.2357, 0.2389, 0.2422, 0.2454, 0.2486,
0.2517, 0.2549, 0.2580, 0.2611, 0.2642, 0.2673, 0.2704, 0.2734, 0.2764, 0.2794,
0.2823, 0.2852, 0.2881, 0.2910, 0.2939, 0.2967, 0.2995, 0.3023, 0.3051, 0.3078,
0.3106, 0.3133, 0.3159, 0.3186, 0.3212, 0.3238, 0.3264, 0.3289, 0.3315, 0.3340,
0.3365, 0.3389, 0.3413, 0.3438, 0.3461, 0.3485, 0.3508, 0.3531, 0.3554, 0.3577,
0.3599, 0.3621, 0.3643, 0.3665, 0.3686, 0.3708, 0.3729, 0.3749, 0.3770, 0.3790,
0.3810, 0.3830, 0.3849, 0.3869, 0.3888, 0.3907, 0.3925, 0.3944, 0.3962, 0.3980,
0.3997, 0.4015, 0.4032, 0.4049, 0.4066, 0.4082, 0.4099, 0.4115, 0.4131, 0.4147,
0.4162, 0.4177, 0.4192, 0.4207, 0.4222, 0.4236, 0.4251, 0.4265, 0.4279, 0.4292,
0.4306, 0.4319, 0.4332, 0.4345, 0.4357, 0.4370, 0.4382, 0.4394, 0.4406, 0.4418,
0.4429, 0.4441, 0.4452, 0.4463, 0.4474, 0.4484, 0.4495, 0.4505, 0.4515, 0.4525,
0.4535, 0.4545, 0.4554, 0.4564, 0.4573, 0.4582, 0.4591, 0.4599, 0.4608, 0.4616,
0.4625, 0.4633, 0.4641, 0.4649, 0.4656, 0.4664, 0.4671, 0.4678, 0.4686, 0.4693,
0.4699, 0.4706, 0.4713, 0.4719, 0.4726, 0.4732, 0.4738, 0.4744, 0.4750, 0.4756,
0.4761, 0.4767, 0.4772, 0.4778, 0.4783, 0.4788, 0.4793, 0.4798, 0.4803, 0.4808,
0.4812, 0.4817, 0.4821, 0.4826, 0.4830, 0.4834, 0.4838, 0.4842, 0.4846, 0.4850,
0.4854, 0.4857, 0.4861, 0.4864, 0.4868, 0.4871, 0.4875, 0.4878, 0.4881, 0.4884,
0.4887, 0.4890, 0.4893, 0.4896, 0.4898, 0.4901, 0.4904, 0.4906, 0.4909, 0.4911,
0.4913, 0.4916, 0.4918, 0.4920, 0.4922, 0.4925, 0.4927, 0.4929, 0.4931, 0.4932,
0.4934, 0.4936, 0.4938, 0.4940, 0.4941, 0.4943, 0.4945, 0.4946, 0.4948, 0.4949,
0.4951, 0.4952, 0.4953, 0.4955, 0.4956, 0.4957, 0.4959, 0.4960, 0.4961, 0.4962,
0.4963, 0.4964, 0.4965, 0.4966, 0.4967, 0.4968, 0.4969, 0.4970, 0.4971, 0.4972,
0.4973, 0.4974, 0.4974, 0.4975, 0.4976, 0.4977, 0.4977, 0.4978, 0.4979, 0.4979,
0.4980, 0.4981, 0.4981, 0.4982, 0.4982, 0.4983, 0.4984, 0.4984, 0.4985, 0.4985,
0.4986, 0.4986, 0.4987, 0.4987, 0.4987, 0.4988, 0.4988, 0.4989, 0.4989, 0.4989,
0.4990, 0.4990]


def sd_to_area(sd):
    sign = 1
    if sd < 0:
        sign = -1
    sd = math.fabs(sd)  # get the absolute value of sd
    index = int(sd * 100)
    if len(area_to_sd_map) <= index:
        return sign * area_to_sd_map[-1] # return last element in array
    if index == (sd * 100):
        return sign * area_to_sd_map[index]
    return sign * (area_to_sd_map[index] + area_to_sd_map[index + 1]) / 2


def area_to_sd(area):
    sign = 1
    if area < 0:
        sign = -1
    area = math.fabs(area)
    for a in range(len(area_to_sd_map)):
        if area == area_to_sd_map[a]:
            return sign * a / 100
        if 0 < a and area_to_sd_map[a - 1] < area and area < area_to_sd_map[a]:
```

```
        # our area is between this value and the previous
        # for simplicity, we will just take the sd half way between a - 1 and a
        return sign * (a - .5) / 100
   return sign * (len(area_to_sd_map) - 1) / 100


def bootstrap(x):
   samp_x = []
   for i in range(len(x)):
     samp_x.append(random.choice(x))
   return samp_x

# subtracts group a mean from group b mean and returns result
def meandiff(grpA, grpB):
   return sum(grpB) / float(len(grpB)) - sum(grpA) / float(len(grpA))


###################################
#
# Computations
#
###################################

# list of lists
samples = []
a = 0
b = 1

# file must be in FASTA format
infile=open(input_file)
for line in infile:
   if line.startswith('>'):
     # start of new sample
     samples.append([])
   elif not line.isspace():
     # line must contain values for previous sample
     samples[len(samples) - 1] += map(float,line.split())
infile.close()

observed_mean_diff = meandiff(samples[a], samples[b])


num_resamples = 10000    # number of times we will resample from our original samples
num_below_observed = 0   # count the number of bootstrap values below the observed
sample statistic
out = []  # will store results of each time we resample

for i in range(num_resamples):
   # get bootstrap samples for each of our groups
   # then compute our statistic of interest
   # append statistic to out
   bootstrap_samples = []  # list of lists
   for sample in samples:
     bootstrap_samples.append(bootstrap(sample))
```

```
  # now we have a list of new samples, run meandiff
  boot_mean_diff = meandiff(bootstrap_samples[a], bootstrap_samples[b])
  if boot_mean_diff < observed_mean_diff:
    num_below_observed += 1
  out.append(boot_mean_diff)

out.sort()

p = num_below_observed / float(num_resamples)# proportion of bootstrap values below
the observed value

dist_from_center = p - .5# if this is negative, the original is below the center, if
positive, it is above
z_0 = area_to_sd(dist_from_center)

# now we want to find the proportion that should be between the mean and one of the
tails
tail_sds = area_to_sd(conf_interval / 2)
z_alpha_over_2 = 0 - tail_sds
z_1_minus_alpha_over_2 = tail_sds

# in case our lower and upper bounds are not integers,
# we decrease the range (the values we include in our interval),
# so that we can keep the same level of confidence
lower_bound = int(math.ceil(num_resamples * (0.5 + sd_to_area(z_alpha_over_2 + (2 *
z_0)))))
upper_bound =  int(math.floor(num_resamples * (0.5 +
sd_to_area(z_1_minus_alpha_over_2 + (2 * z_0)))))


###################################
#
# Output
#
###################################

print "Observed difference between the means: %.2f" % observed_mean_diff
print "We have", conf_interval * 100, "% confidence that the true difference between
the means",
print "is between: %.2f" % out[lower_bound], "and %.2f" % out[upper_bound]
```

MeanConf.py

```
#!/usr/bin/python

###################################
# Mean Confidence Interval
# From: Statistics is Easy! By Dennis Shasha and Manda Wilson
#
#
# Uses shuffling & bootstrapping to get a 90% confidence interval for the mean.
#
```

```
# Author: Manda Wilson
#
# Example of FASTA formatted input file:
# >slides
# 60.2 63.1 58.4 58.9 61.2 67.0 61.0 59.7 58.2 59.8
#
# Included in the code, but NOT in the pseudocode,
# is the bias-corrected confidence interval.
# See http://www.statisticsiseasy.com/BiasCorrectedConfInter.html
# for more information on bias-corrected cofidence intervals.
#
# Pseudocode:
#
# 1. Measure the mean.  The mean is computed by by sum(grp) / len(grp).
#     In this example the difference between the two group means is 12.97.
#
# 2. Do the following 10,000 times:
#     a. Create a new array of the same size as the original sample
#     b. Fill the array with randomly picked values from the original sample (randomly
picked with replacement)
#     b. Measure the mean, just as we did in step (1) with the original sample.
#
# 3. Sort the means computed in step (2).
#
# 4. Compute the size of each interval tail.  If we want a 90% confidence interval,
then 1 - 0.9 yields the
#    portion of the interval in the tails.  We divide this by 2 to get the size of each
tail, in this case 0.05.
#
# 5. Compute the upper and lower bounds.  To get the lower bound we multiply the tail
size by the number of
#    bootstraps we ran and round this value up to the nearest integer (to make sure
this value maps
#    to an actual boostrap we ran).  In this case we multiple 0.05 by 10,000 and get
500, which rounds up to 500.
#
#    To compute the upper bound we subtract the tail size from 1 and then multiply that
by the number of bootstraps
#    we ran.  Round the result of the last step down to the nearest integer.  Again,
this is to ensure this value
#    maps to an actual bootstrap we ran.  We round the lower bound up and the upper
bound down to reduce the
#    confidence interval size, so that we can still say we have as much confidence in
this result.
#
# 6. The bootstrap values at the lower bound and upper bound give us our confidence
interval.
#
####################################


import random
import math
```

```
import sys


####################################
#
# Adjustable variables
#
####################################


input_file = "input/MeanConf.vals"
conf_interval = 0.9


####################################
#
# Subroutines
#
####################################


# maps proportion of values above mean
# to number of standard deviations above mean
# keys will be index / 100 \:[0-9]\.[0-9][0-9]\,
area_to_sd_map = [0.0000, 0.0040, 0.0080, 0.0120, 0.0160, 0.0199, 0.0239, 0.0279,
    0.0319, 0.0359, 0.0398, 0.0438, 0.0478, 0.0517, 0.0557, 0.0596, 0.0636, 0.0675,
    0.0714, 0.0753, 0.0793, 0.0832, 0.0871, 0.0910, 0.0948, 0.0987, 0.1026, 0.1064,
    0.1103, 0.1141, 0.1179, 0.1217, 0.1255, 0.1293, 0.1331, 0.1368, 0.1406, 0.1443,
    0.1480, 0.1517, 0.1554, 0.1591, 0.1628, 0.1664, 0.1700, 0.1736, 0.1772, 0.1808,
    0.1844, 0.1879, 0.1915, 0.1950, 0.1985, 0.2019, 0.2054, 0.2088, 0.2123, 0.2157,
    0.2190, 0.2224, 0.2257, 0.2291, 0.2324, 0.2357, 0.2389, 0.2422, 0.2454, 0.2486,
    0.2517, 0.2549, 0.2580, 0.2611, 0.2642, 0.2673, 0.2704, 0.2734, 0.2764, 0.2794,
    0.2823, 0.2852, 0.2881, 0.2910, 0.2939, 0.2967, 0.2995, 0.3023, 0.3051, 0.3078,
    0.3106, 0.3133, 0.3159, 0.3186, 0.3212, 0.3238, 0.3264, 0.3289, 0.3315, 0.3340,
    0.3365, 0.3389, 0.3413, 0.3438, 0.3461, 0.3485, 0.3508, 0.3531, 0.3554, 0.3577,
    0.3599, 0.3621, 0.3643, 0.3665, 0.3686, 0.3708, 0.3729, 0.3749, 0.3770, 0.3790,
    0.3810, 0.3830, 0.3849, 0.3869, 0.3888, 0.3907, 0.3925, 0.3944, 0.3962, 0.3980,
    0.3997, 0.4015, 0.4032, 0.4049, 0.4066, 0.4082, 0.4099, 0.4115, 0.4131, 0.4147,
    0.4162, 0.4177, 0.4192, 0.4207, 0.4222, 0.4236, 0.4251, 0.4265, 0.4279, 0.4292,
    0.4306, 0.4319, 0.4332, 0.4345, 0.4357, 0.4370, 0.4382, 0.4394, 0.4406, 0.4418,
    0.4429, 0.4441, 0.4452, 0.4463, 0.4474, 0.4484, 0.4495, 0.4505, 0.4515, 0.4525,
    0.4535, 0.4545, 0.4554, 0.4564, 0.4573, 0.4582, 0.4591, 0.4599, 0.4608, 0.4616,
    0.4625, 0.4633, 0.4641, 0.4649, 0.4656, 0.4664, 0.4671, 0.4678, 0.4686, 0.4693,
    0.4699, 0.4706, 0.4713, 0.4719, 0.4726, 0.4732, 0.4738, 0.4744, 0.4750, 0.4756,
    0.4761, 0.4767, 0.4772, 0.4778, 0.4783, 0.4788, 0.4793, 0.4798, 0.4803, 0.4808,
    0.4812, 0.4817, 0.4821, 0.4826, 0.4830, 0.4834, 0.4838, 0.4842, 0.4846, 0.4850,
    0.4854, 0.4857, 0.4861, 0.4864, 0.4868, 0.4871, 0.4875, 0.4878, 0.4881, 0.4884,
    0.4887, 0.4890, 0.4893, 0.4896, 0.4898, 0.4901, 0.4904, 0.4906, 0.4909, 0.4911,
    0.4913, 0.4916, 0.4918, 0.4920, 0.4922, 0.4925, 0.4927, 0.4929, 0.4931, 0.4932,
    0.4934, 0.4936, 0.4938, 0.4940, 0.4941, 0.4943, 0.4945, 0.4946, 0.4948, 0.4949,
    0.4951, 0.4952, 0.4953, 0.4955, 0.4956, 0.4957, 0.4959, 0.4960, 0.4961, 0.4962,
    0.4963, 0.4964, 0.4965, 0.4966, 0.4967, 0.4968, 0.4969, 0.4970, 0.4971, 0.4972,
    0.4973, 0.4974, 0.4974, 0.4975, 0.4976, 0.4977, 0.4977, 0.4978, 0.4979, 0.4979,
    0.4980, 0.4981, 0.4981, 0.4982, 0.4982, 0.4983, 0.4984, 0.4984, 0.4985, 0.4985,
    0.4986, 0.4986, 0.4987, 0.4987, 0.4987, 0.4988, 0.4988, 0.4989, 0.4989, 0.4989,
    0.4990, 0.4990]


def sd_to_area(sd):
```

```
    sign = 1
    if sd < 0:
       sign = -1
    sd = math.fabs(sd)   # get the absolute value of sd
    index = int(sd * 100)
    if len(area_to_sd_map) <= index:
       return sign * area_to_sd_map[-1] # return last element in array
    if index == (sd * 100):
       return sign * area_to_sd_map[index]
    return sign * (area_to_sd_map[index] + area_to_sd_map[index + 1]) / 2

def area_to_sd(area):
   sign = 1
   if area < 0:
      sign = -1
   area = math.fabs(area)
   for a in range(len(area_to_sd_map)):
      if area == area_to_sd_map[a]:
         return sign * a / 100
      if 0 < a and area_to_sd_map[a - 1] < area and area < area_to_sd_map[a]:
         # our area is between this value and the previous
         # for simplicity, we will just take the sd half way between a - 1 and a
         return sign * (a - .5) / 100
   return sign * (len(area_to_sd_map) - 1) / 100

def bootstrap(x):
        samp_x = []
        for i in range(len(x)):
                samp_x.append(random.choice(x))
        return samp_x


# expects list containing one list
def mean(grp):
        return sum(grp) / float(len(grp))

###################################
#
# Computations
#
###################################

sample = []

# file must be in FASTA format
infile=open(input_file)
for line in infile:
        if not line.isspace() and not line.startswith('>'):
                # line must contain values for previous sample
                sample += map(float,line.split())
infile.close()
```

```
observed_mean = mean(sample)

num_resamples = 10000    # number of times we will resample from our original samples
num_below_observed = 0    # count the number of bootstrap values below the observed
sample statistic
out = []                  # will store results of each time we resample

for i in range(num_resamples):
  # get bootstrap sample
  # then compute mean
  # append mean to out
  boot_mean = mean(bootstrap(sample))
  if boot_mean < observed_mean:
    num_below_observed += 1
  out.append(boot_mean)

out.sort()

# standard confidence interval computations
tails = (1 - conf_interval) / 2

# in case our lower and upper bounds are not integers,
# we decrease the range (the values we include in our interval),
# so that we can keep the same level of confidence
lower_bound = int(math.ceil(num_resamples * tails))
upper_bound = int(math.floor(num_resamples * (1 - tails)))

# bias-corrected confidence interval computations
p = num_below_observed / float(num_resamples)# proportion of bootstrap values below
the observed value

dist_from_center = p - .5# if this is negative, the original is below the center, if
positive, it is above
z_0 = area_to_sd(dist_from_center)

# now we want to find the proportion that should be between the mean and one of the
tails
tail_sds = area_to_sd(conf_interval / 2)
z_alpha_over_2 = 0 - tail_sds
z_1_minus_alpha_over_2 = tail_sds

# in case our lower and upper bounds are not integers,
# we decrease the range (the values we include in our interval),
# so that we can keep the same level of confidence
bias_corr_lower_bound = int(math.ceil(num_resamples * (0.5 +
sd_to_area(z_alpha_over_2 + (2 * z_0)))))
bias_corr_upper_bound =  int(math.floor(num_resamples * (0.5 +
sd_to_area(z_1_minus_alpha_over_2 + (2 * z_0)))))


###################################
#
```

```
# Output
#
#####################################

# print observed value and then confidence interval
print "Observed mean: %.2f" % observed_mean
print "We have", conf_interval * 100, "% confidence that the true mean",
print "is between: %.2f" % out[lower_bound], "and %.2f" % out[upper_bound]

print "***** Bias Corrected Confidence Interval *****"
print "We have", conf_interval * 100, "% confidence that the true mean",
print "is between: %.2f" % out[bias_corr_lower_bound], "and %.2f" %
out[bias_corr_upper_bound]
```

MeanConf.vals

```
>slides
60.2 63.1 58.4 58.9 61.2 67.0 61.0 59.7 58.2 59.8
```

ChiSquaredOne.py

```
#!/usr/bin/python

#####################################
# One Variable Chi-Squared Significance Test
# From: Statistics is Easy! By Dennis Shasha and Manda Wilson
#
#
# Simulates 10,000 experiments, where each experiment is 60 tosses of a fair die.
# For each experiment we count the number of 1s, 2s, 3s, 4s, 5s, and 6s that occur.
# We then compare our observed number of  1s, 2s, 3s, 4s, 5s, and 6s in 60 tosses
# of a real die to this distribution, to determine the probability that our sample
# came from this assumed distribution (to decide whether this die is fair).
#
# Author: Manda Wilson
#
# Example of FASTA formatted input file:
# >expected
# 10 10 10 10 10 10
# >observed
# 14 16 6 9 5 10
#
# Pseudocode:
#
# 1. Calculate chi-squared for the observed values (in our example it is 9.4).
#      a. For each category (in this example 1, 2, 3, 4, 5, and 6 are the categories):
#          i. Subtract the expected count from the observed count
#          ii. Square the result of step (1ai)
#          iii. Divide the result of step (1aii) by the expected count
#      b. Sum the results of step (1a), this is r, our observed chi-squared value
#
# 2. Set a counter to 0, this will count the number of times we get a chi-squared
#      greater than or equal to 9.4.
```

```
#
# 3. Do the following 10,000 times:
#    a. Create an array that is equal in length to the number of categories (in our
example, 6).
#       All values in this array should start as 0.  This array will be used to count
the number
#       of observations for each category (the number of times we roll a 1, a 2, etc.).
#    b. Use the total sum of expected counts to determine how many times to do the
following (in
#       our example there are a total of 60 die rolls, so we do this 60 times):
#       i. Pick a random number and increment the count in the bin it corresponds to
#    c. Calculate chi-squared on the results from step (3b), just as we did in step
(1).
#    d. If the result from step (3c) is greater than or equal to our observed chi-
squared (9.4), increment our counter
#       from step (2).
#
# 3. counter / 10,000 equals the probability of getting a chi-squared greater than or
equal to
#    9.4, if the die is in fact fair.
#
#####################################

import random

#####################################
#
# Adjustable variables
#
#####################################

#####################################
#
# Subroutines
#
#####################################

# will draw num values, given
# probabilities given by counts in expected
def drawfromcategories(num, expected):
    # if we have two bins (two categories)
    # bin 1 and bin 2
    # and we expect bin 1 to get a hit 2 of 5 times
    # and bin 2 to get a hit 3 of 5
    # then we store 2 in bins[0], and 5 in bins[1]
    # then if we draw anything 2 or under we know it is for bin 1
    # else if we draw anything 5 or under we know it is for bin 2
    # the expected values give us our probabilities
    max = 0
    bins = []
    observed = []
    # in this loop we weight each bin
```

```
    # and we initialize observed counts to 0
    for b in range(len(expected)):
       max += expected[b]
       bins.append(max)
       observed.append(0)
    num_bins = len(bins)
    for d in range(num):
       draw = random.randint(1, max)
       # which bin does this belong in?
       b = 0
       while b < num_bins and draw > bins[b]:
          b += 1# move to next bin
       observed[b] += 1 # this is the category that was drawn
    return observed

def chisquared(expected, observed):
   count = len(expected)
   total = 0
   for i in range (count):
      total += ((observed[i] - expected[i])**2) / float(expected[i])
   return total


###################################
#
# Computations
#
###################################

# list of lists
samples = []

# file must be in FASTA format
infile=open('input/ChiSquared.vals')
for line in infile:
        if line.startswith('>'):
                # start of new sample
                samples.append([])
        elif not line.isspace():
                # line must contain values for previous sample
                samples[len(samples) - 1] += map(float,line.split())
infile.close()

expected = samples[0]
observed = samples[1]

observed_chi_squared = chisquared(expected, observed)

num_observations = int(sum(observed))

count = 0
num_runs = 10000
```

```
for i in range(num_runs):
    # roll a fair die num_observations times, counting the results
    simulated_observed = drawfromcategories(num_observations, expected)
    chi_squared = chisquared(expected, simulated_observed)
    if (chi_squared >= observed_chi_squared):
        count = count + 1


#####################################
#
# Output
#
#####################################

print "Observed chi-squared: %.2f" % observed_chi_squared
print count, "out of 10000 experiments had a chi-squared difference greater than or
equal to %.2f" % observed_chi_squared
print "Probability that chance alone gave us a chi-squared greater than or equal to",
print "%.2f" % observed_chi_squared, "is", (count / float(num_runs))
```

ChiSquared.vals

```
>expected
10 10 10 10 10 10
>observed
14 16 6 9 5 10
```

ChiSquaredMulti.py

```
#!/usr/bin/python

#####################################
# One Variable Chi-Squared Significance Test
# From: Statistics is Easy! By Dennis Shasha and Manda Wilson
#
#
# Simulates 10,000 experiments, where each experiment is 60 tosses of a fair die.
# For each experiment we count the number of 1s, 2s, 3s, 4s, 5s, and 6s that occur.
# We then compare our observed number of  1s, 2s, 3s, 4s, 5s, and 6s in 60 tosses
# of a real die to this distribution, to determine the probability that our sample
# came from this assumed distribution (to decide whether this die is fair).
#
# Author: Manda Wilson
#
# Example of FASTA formatted input file:
# >expected
# 10 10 10 10 10 10
# >observed
# 14 16 6 9 5 10
#
# Pseudocode:
#
# 1. Calculate chi-squared for the observed values (in our example it is 9.4).
```

```
#      a. For each category (in this example 1, 2, 3, 4, 5, and 6 are the categories):
#         i. Subtract the expected count from the observed count
#         ii. Square the result of step (1ai)
#         iii. Divide the result of step (1aii) by the expected count
#      b. Sum the results of step (1a), this is r, our observed chi-squared value
#
# 2. Set a counter to 0, this will count the number of times we get a chi-squared
#    greater than or equal to 9.4.
#
# 3. Do the following 10,000 times:
#      a. Create an array that is equal in length to the number of categories (in our
example, 6).
#         All values in this array should start as 0.  This array will be used to count
the number
#         of observations for each category (the number of times we roll a 1, a 2, etc.).
#      b. Use the total sum of expected counts to determine how many times to do the
following (in
#         our example there are a total of 60 die rolls, so we do this 60 times):
#         i. Pick a random number and increment the count in the bin it corresponds to
#      c. Calculate chi-squared on the results from step (3b), just as we did in step
(1).
#      d. If the result from step (3c) is greater than or equal to our observed chi-
squared (9.4), increment our counter
#         from step (2).
#
# 3. counter / 10,000 equals the probability of getting a chi-squared greater than or
equal to
#    9.4, if the die is in fact fair.
#
#####################################

import random

#####################################
#
# Adjustable variables
#
#####################################

#####################################
#
# Subroutines
#
#####################################

# will draw num values, given
# probabilities given by counts in expected
def drawfromcategories(num, expected):
    # if we have two bins (two categories)
    # bin 1 and bin 2
    # and we expect bin 1 to get a hit 2 of 5 times
    # and bin 2 to get a hit 3 of 5
```

```
    # then we store 2 in bins[0], and 5 in bins[1]
    # then if we draw anything 2 or under we know it is for bin 1
    # else if we draw anything 5 or under we know it is for bin 2
    # the expected values give us our probabilities
    max = 0
    bins = []
    observed = []
    # in this loop we weight each bin
    # and we initialize observed counts to 0
    for b in range(len(expected)):
       max += expected[b]
       bins.append(max)
       observed.append(0)
    num_bins = len(bins)
    for d in range(num):
       draw = random.randint(1, max)
       # which bin does this belong in?
       b = 0
       while b < num_bins and draw > bins[b]:
          b += 1# move to next bin
       observed[b] += 1 # this is the category that was drawn
    return observed

def chisquared(expected, observed):
    count = len(expected)
    total = 0
    for i in range (count):
       total += ((observed[i] - expected[i])**2) / float(expected[i])
    return total


###################################
#
# Computations
#
###################################

# list of lists
samples = []

# file must be in FASTA format
infile=open('input/ChiSquared.vals')
for line in infile:
        if line.startswith('>'):
                # start of new sample
                samples.append([])
        elif not line.isspace():
                # line must contain values for previous sample
                samples[len(samples) - 1] += map(float,line.split())
infile.close()

expected = samples[0]
observed = samples[1]
```

```
observed_chi_squared = chisquared(expected, observed)

num_observations = int(sum(observed))

count = 0
num_runs = 10000

for i in range(num_runs):
  # roll a fair die num_observations times, counting the results
  simulated_observed = drawfromcategories(num_observations, expected)
  chi_squared = chisquared(expected, simulated_observed)
  if (chi_squared >= observed_chi_squared):
    count = count + 1

#####################################
#
# Output
#
#####################################

print "Observed chi-squared: %.2f" % observed_chi_squared
print count, "out of 10000 experiments had a chi-squared difference greater than or
equal to %.2f" % observed_chi_squared
print "Probability that chance alone gave us a chi-squared greater than or equal to",
print "%.2f" % observed_chi_squared, "is", (count / float(num_runs))
```

ChiSquaredmulti.vals

```
>sick
20 18 8
>healthy
24 24 16
```

FishersExactTestSig.py

```
#!/usr/bin/python

#####################################
# Fisher's Exact Test - Significance Test
# From: Statistics is Easy! By Dennis Shasha and Manda Wilson
#
#
# Assuming that our tea taster can not identify which cups of tea were milk
# first and which were not, tests to see the probability of getting an outcome
# as extreme or more extreme than the observed one (i.e. more correct than in the
# observed case) by chance alone.
#
# This is a one-tailed test.
#
# Author: Manda Wilson
#
```

```
# Example of FASTA formatted input file:
# >tea taster claimed milk first
# 3 1
# >tea taster claimed tea first
# 2 4
#
# The above generalizes to:
#
# >variable 1
# a b
# >variable 2
# c d
#
# Pseudocode:
#
# 1. Calculate Fisher's Exact Test for the observed values (in our example it is
0.2619).
#     a. Calculate a + b, c + d, a + c, b + d, and a + b + c + d
#     b. Set observed_prob_tail = 0
#     c. Compute all possible outcomes (matricies)
#        i. For a' = 0 to a + b + c + d
#            A. b' = (a + b) - a' (if this is impossible, skip to next a')
#            B. c' = (a + c) - a' (if this is impossible, skip to next a')
#            C. d' = (c + d) - c' (if this is impossible, skip to next a')
#            D. If this matrix is "as extreme", or "more extreme" than the observed
compute the
#                probability of getting this matrix by chance and add this probability
#                to observed_prob_tail. In this case, the matrix is "more extreme" if
#                a' + d' > a + b, i.e., the tea taster was correct more often than
observed.
#                If the test is a two-tailed test, "more extreme" matricies include
#                those where a' + d' > a + b as well as those where b' + c' > a + b
#                (i.e. the matrix is more unbalanced than the observed).
#                - Compute the probability of getting the matrix
#                  a' b'
#                  c' d'
#                  prob = ((a' + b')!(c' + d')!(a' + c')!(b' + d')! / (a'!b'!c'!d'!n'!)
#                  Where n = a' + b' + c' + d'
#                  Example: ((3 + 1)!(2 + 4)!(3 + 2)!(1 + 4)!) / (3!1!2!4!10!) = 10/42
#                - observed_prob_tail += probability from step (1ciD)
#
# 2. Set a counter to 0, this will count the number of times we get a fisher's exact
test
#     less than or equal to 0.2619 (less than because a smaller value is more
unlikely).
#
# 3. Do the following 10,000 times:
#     a. Create a new matrix of counts, preserving the marginals from our original
matrix
#        i.  While there are more rows:
#            I. While there are more columns:
#                * If we ARE NOT at the last element in the row
```

```
#                 * If we ARE NOT at the last element in the column
#                    - Pick a random number between 0
#                       and min(available_in_this_row,
#                          available_in_this_column) (inclusive)
#                    - Store this value in our new maxtrix in the
#                       current row, current column position
#                 * If we ARE at the last element in the column
#                    - Store whatever is available_in_this_column
#                       in the current row, current column position
#                 * If we ARE at the last element in the row
#                    - Store whatever is available_in_this_row
#                       in the current row, current column position
#                 * Subtract whatever is stored in the current row,
#                    current column position from the available_in_this_column
#                    as well as from available_in_this_row
#
#    b. Calculate Fisher's Exact Test on the results from step (3a), just as we did in
step (1).
#    c. If the result from step (3b) is less than or equal to our observed Fisher's
Exact Test (0.2619),
#       increment our counter from step (2).
#
# 4. counter / 10,000 equals the probability of getting a Fisher's Exact Test less
than
#    or equal to our observed probability (0.2619)
#
###################################

import random

###################################
#
# Adjustable variables
#
###################################

input_file = "input/fishersexact.vals"

###################################
#
# Subroutines
#
###################################

# takes a list of values, plus row and column totals
# For an m x n matrix, values must be ordered:
# row 1, columns 1 - n, row 2, columns 1 - n, ... row m, columns 1 - n
# shuffles values, preserving row and column totals
def shuffle(orig_row_totals, orig_column_totals):
        # start with an empty m x n matrix
        # fill it with randomly generated counts
        # that preserve the row and column totals
```

```
        current_row = 0
        # keeps track of the available values for this row
        available_row_vals = orig_row_totals[:]
        # keeps track of the available values for this column
        available_column_vals = orig_column_totals[:]

        new_counts = []

        while current_row < len(orig_row_totals):
                current_column = 0
                while current_column < len(orig_column_totals):
                        # if we are not at the last element in either the row or the
column
                        if current_row < len(orig_row_totals) - 1:
                                if current_column < len(orig_column_totals) - 1:
                                        # get a random number between 0 and
                                        # min(available_row_vals[current_row],
                                        #
available_column_vals[current_column])
                                        max_val = min(available_row_vals[current_row],
available_column_vals[current_column])
                                        new_val = random.randint(0, max_val)
                                        # put this value in the new matrix
                                        new_counts.append(new_val)
                                else: # we are at the last column, this value must be
whatever is available for this row

new_counts.append(available_row_vals[current_row])
                        else: # we are at the last row, this value must be whatever is
available for this column
                                new_counts.append(available_column_vals[current_column])
                        # remove this amount from both the available row and column
values
                        available_row_vals[current_row] -= new_counts[-1]
                        available_column_vals[current_column] -= new_counts[-1]
                        current_column += 1
                current_row += 1
        return new_counts

def factorial(n):
  prod = 1
  for i in range(n):
    prod = prod * (i + 1)
  return prod

def prob_of_matrix(a, b, c, d):
  return (factorial(a + b) * factorial(c + d) * factorial(a + c) * factorial(b + d))
/ float(factorial(a) * factorial(b) * factorial(c) * factorial(d) * factorial(a + b +
c + d))

def fishers_exact_test(a, b, c, d):
  # now we have to figure out possible outcomes
```

```
# that are more extreme than ours
# and sum the probability of each
# this is the part of the code that should be tailored
# for your definition of "more extreme"
# here we are doing a one-tailed test
# where "more extreme" means, more correct answers than
# what was observed
# this translates to any matrix where a + d is larger than ours

prob_tail = 0.0

for a_prime in range(a + b + c + d + 1):
      b_prime = a + b - a_prime
      if b_prime >= 0:
              c_prime = a + c - a_prime
              if c_prime >= 0:
                      d_prime = c + d - c_prime
                      if d_prime >= 0:
                              # this matrix is valid
                              # now check if it is "more extreme" than the observed
                              if a_prime + d_prime >= a + d:
                                      prob_tail = prob_tail +
prob_of_matrix(a_prime, b_prime, c_prime, d_prime)
   return prob_tail


###################################
#
# Computations
#
###################################

# set to invalid counts
a = -1
b = -1
c = -1
d = -1

# file must be in FASTA format
infile=open(input_file)
for line in infile:
   if not line.isspace() and not line.startswith('>'):
     # this is one row
     if a == -1 and b == -1:
        (a, b) = map(int,line.split())
     else:
        (c, d) = map(int,line.split())
infile.close()

observed_prob_tail = fishers_exact_test(a, b, c, d)

count = 0
num_runs = 10000
```

APPENDIX B   109

```
for i in range(num_runs):
        [a_prime, b_prime, c_prime, d_prime] = shuffle([a + b, c + d], [a + c, b + d])
        prob_tail = fishers_exact_test(a_prime, b_prime, c_prime, d_prime)
        if (prob_tail <= observed_prob_tail):
                count = count + 1

#####################################
#
# Output
#
#####################################

print "Observed Fisher's Exact Test: %.4f" % observed_prob_tail
print count, "out of 10000 experiments had a Fisher's Exact Test less than or equal to
%.4f" % observed_prob_tail
print "Probability that chance alone gave us a Fisher's Exact Test",
print "of %.4f" % observed_prob_tail, "or less is", (count / float(num_runs))
```

fishersexact.vals

```
>tea taster claimed milk first
3 1
>tea taster claimed tea first
2 4
```

OneWayAnovaSig.py

```
#!/usr/bin/python

#####################################
# One-Way ANOVA Significance Test
# From: Statistics is Easy! By Dennis Shasha and Manda Wilson
#
#
# Assuming that there is no difference in the three drugs used, tests to see the
probability
# of getting a f-statistic by chance alone greater than or equal to the one observed.
# Uses shuffling to get a distribution to compare
# to the observed f-statistic.
#
# Author: Manda Wilson
#
# Example of FASTA formatted input file:
# >grp_a
# 45 44 34 33 45 46 34
# >grp_b
# 34 34 50 49 48 39 45
# >grp_c
# 24 34 23 25 36 28 33 29
#
# Pseudocode:
```

```
#
# 1. Calculate f-statistic for the observed values (in our example it is 11.27).
#     a. Initialize within sum of squares (wss), total sum (ts),
#        total count (tc), and between sum of squares (bss) to 0
#     b. For each group:
# i. Within group mean (wgm) = sum of group values / number of values in group
# ii. Store the number of values in group
# iii. Sum the following: for each value in the group
#           I. Subtract the value from wgm
#           II. Square the result of step (1biiiI)
#           III.  Add the result of step (1biiiII) to wss
#     c. Total mean (tm) = ts / tc
#     d. For each group:
#        i. Subtract the wgm from tm
#        ii. Square the result of step (1di)
#        iii. Multiply the result of step (1dii) by
#             the number of values in that group (stored in step (1bii))
#        iv. Add the result of step (1diii) to bss
#     e. Between degrees of freedom (bdf) = number of groups - 1
#     f. Within degrees of freedome (wdf) = tc - number of groups
#     g. Within group variance (wgv) = wss / wdf
#     h. Between group variance (bgv) = bss / bdf
#     i. f-statistic = wgv / bgv
#
# 2. Set a counter to 0, this will count the number of times we get a f-statistic
#     greater than or equal to 11.27.
#
# 3. Do the following 10,000 times:
#     a. Shuffle the observed values. To do this:
#        i. Put the values from all the groups into one array
#        ii. Shuffle the pooled values
#        iii. Reassign the pooled values to groups of the same size as the original
groups
#     b. Calculate f-statistic on the results from step (3a), just as we did in step (1)
#     c. If the result from step (3b) is greater than or equal to our observed f-
statistic (11.27),
#           increment our counter from step (2).
#
# 3. counter / 10,000 equals the probability of getting a f-statistic greater than or
equal to
#     11.27, assuming there is no difference between the groups
#
#####################################

import random

#####################################
#
# Adjustable variables
#
#####################################
```

```python
input_file = 'input/OneWayAnova.vals'


####################################
#
# Subroutines
#
####################################


# a list of lists
def shuffle(grps):
  num_grps = len(grps)
  pool = []

  # throw all values together
  for i in range(num_grps):
    pool.extend(grps[i])
  # mix them up
  random.shuffle(pool)
  # reassign to groups
  new_grps = []
  start_index = 0
  end_index = 0
  for i in range(num_grps):
    end_index = start_index + len(grps[i])
    new_grps.append(pool[start_index:end_index])
    start_index = end_index
  return new_grps


def sumofsq(vals, mean):
  # the sum of squares for a group is calculated
  # by getting the sum of the squared difference
  # of each value and the mean of the group that value belongs to
  count = len(vals)
  total = 0
  for i in range (count):
    diff_sq = (vals[i] - mean)**2
    total += diff_sq
  return total

def weightedsumofsq(vals, weights, mean):
  count = len(vals)
  total = 0
  for i in range (count):
    diff_sq = (vals[i] - mean)**2
    total += weights[i] * diff_sq
  return total

# expects list of lists
def onewayanova(grps):
  num_grps = len(grps)
  within_group_means = []
```

```
grp_counts = []
within_ss = 0
total_sum = 0
total_count = 0
for i in range (num_grps):
    grp = grps[i]
    grp_count = len(grp)
    grp_sum = sum(grp)
    within_group_mean = grp_sum / float(grp_count)

    grp_counts.append(grp_count)

    total_count += grp_count# add to total number of vals
    total_sum += grp_sum

    within_group_means.append(within_group_mean)

    # to get the within group sum of squares:
    # sum the following: for every element in the overall group
    # subtract that element's value from that element's group mean
    # square the difference
    # get within group sum of squares
    # this is calculated by summing each group's sum of squares
    within_ss += sumofsq(grp, within_group_mean)

total_mean = total_sum / total_count

# to get the between group sum of squares:
# sum the following: for every element in the overall group
# subtract that element's group mean from the overall group mean
# square the difference
# grp_counts are used as weights
between_ss = weightedsumofsq(within_group_means, grp_counts, total_mean)

# now we want to find out how different the groups are between each other
# compared to how much the values vary within the groups
# if all groups vary a lot within themselves,
# and there is no significant difference
# between the groups, then we expect the differences between
# the groups to vary by about the same amount
# so lets get the ratio of the between group variance and
# the within group variance
# if the ratio is 1, then there is no difference between the groups
# if it is significantly larger than one,
# then there is a significant difference between the groups
# remember: even if the groups are significantly different,
# we still won't know which groups are different

# the between group degrees of freedom
# is equal to the number of groups - 1
# this is because once we know the number of groups - 1,
# we know the last group
```

```
    between_df = len(grp_counts) - 1

    # the within group degrees of freedom
    # is equal to the total number of values minus the number of groups
    # this is because for each group, once we know the count - 1 values,
    # we know the last value for that group
    # so we lose the number of groups * 1 degrees of freedom
    within_df = total_count - num_grps

    within_var = within_ss / within_df
    between_var = between_ss / between_df

    f_stat = between_var / within_var

    return f_stat


######################################
#
# Computations
#
######################################

# list of lists
samples = []

# file must be in FASTA format
infile=open(input_file)
for line in infile:
        if line.startswith('>'):
                # start of new sample
                samples.append([])
        elif not line.isspace():
                # line must contain values for previous sample
                samples[len(samples) - 1] += map(float,line.split())
infile.close()

observed_f_statistic = onewayanova(samples)

count = 0
num_shuffles = 10000

for i in range(num_shuffles):
  new_samples = shuffle(samples)
  f_statistic = onewayanova(new_samples)
  if (f_statistic >= observed_f_statistic):
    count = count + 1


######################################
#
# Output
#
######################################
```

```
print "Observed F-statistic: %.2f" % observed_f_statistic
print count, "out of 10000 experiments had a F-statistic greater than or equal to
%.2f" % observed_f_statistic
print "Probability that chance alone gave us a F-statistic",
print "of %.2f or more" % observed_f_statistic, "is", (count / float(num_shuffles))
```

OneWayAnova.vals

```
>grp_a
45 44 34 33 45 46 34
>grp_b
34 34 50 49 48 39 45
>grp_c
24 34 23 25 36 28 33 29
```

OneWayAnovaConf.py

```
#!/usr/bin/python

######################################
# One-Way ANOVA Confidence Interval
# From: Statistics is Easy! By Dennis Shasha and Manda Wilson
#
#
# Uses shuffling & bootstrapping to get a 90% confidence interval for the f-statistic.
#
# Author: Manda Wilson
#
# Example of FASTA formatted input file:
# >grp_a
# 45 44 34 33 45 46 34
# >grp_b
# 34 34 50 49 48 39 45
# >grp_c
# 24 34 23 25 36 28 33 29
#
# Included in the code, but NOT in the pseudocode,
# is the bias-corrected confidence interval.
# See http://www.statisticsiseasy.com/BiasCorrectedConfInter.html
# for more information on bias-corrected cofidence intervals.
#
# Pseudocode:
#
# 1. Calculate f-statistic for the observed values (in our example it is 11.27).
#     a. Initialize within sum of squares (wss), total sum (ts),
#         total count (tc), and between sum of squares (bss) to 0
#     b. For each group:
#         i. Within group mean (wgm) = sum of group values / number of values in group
#         ii. Store the number of values in group
#         iii. Sum the following: for each value in the group
#             I. Subtract the value from wgm
#             II. Square the result of step (1biiiI)
```

```
#            III.  Add the result of step (1biiiII) to wss
#    c. Total mean (tm) = ts / tc
#    d. For each group:
#        i. Subtract the wgm from tm
#        ii. Square the result of step (1di)
#        iii. Multiply the result of step (1dii) by
#             the number of values in that group (stored in step (1bii))
#        iv. Add the result of step (1diii) to bss
#    e. Between degrees of freedom (bdf) = number of groups - 1
#    f. Within degrees of freedome (wdf) = tc - number of groups
#    g. Within group variance (wgv) = wss / wdf
#    h. Between group variance (bgv) = bss / bdf
#    i. f-statistic = wgv / bgv
#
# 2. Do the following 10,000 times:
#    a. For each sample we have get a bootstrap sample:
#        i. Create a new array of the same size as the original sample
#        ii. Fill the array with randomly picked values from the original sample
(randomly picked with replacement)
#    b. Calculate the f-statistic on the bootstrap samples, just as we did in step (1)
#        with the original samples.
#
# 3. Sort the f-statistics computed in step (2).
#
# 4. Compute the size of each interval tail.  If we want a 90% confidence interval,
then 1 - 0.9 yields the
#    portion of the interval in the tails.  We divide this by 2 to get the size of each
tail, in this case 0.05.
#
# 5. Compute the upper and lower bounds.  To get the lower bound we multiply the tail
size by the number of
#    bootstraps we ran and round this value up to the nearest integer (to make sure
this value maps
#    to an actual boostrap we ran).  In this case we multiple 0.05 by 10,000 and get
500, which rounds up to 500.
#
#    To compute the upper bound we subtract the tail size from 1 and then multiply that
by the number of bootstraps
#    we ran.  Round the result of the last step down to the nearest integer.  Again,
this is to ensure this value
#    maps to an actual bootstrap we ran.  We round the lower bound up and the upper
bound down to reduce the
#    confidence interval size, so that we can still say we have as much confidence in
this result.
#
# 6. The bootstrap values at the lower bound and upper bound give us our confidence
interval.
#
####################################

import random
import math
import sys
```

```
#####################################
#
# Adjustable variables
#
#####################################

input_file = 'input/OneWayAnova.vals'

#####################################
#
# Subroutines
#
#####################################

# maps proportion of values above statistic
# to number of standard deviations above statistic
# keys will be index / 100 \:[0-9]\.[0-9][0-9]\,
area_to_sd_map = [0.0000, 0.0040, 0.0080, 0.0120, 0.0160, 0.0199, 0.0239, 0.0279,
0.0319, 0.0359, 0.0398, 0.0438, 0.0478, 0.0517, 0.0557, 0.0596, 0.0636, 0.0675,
0.0714, 0.0753, 0.0793, 0.0832, 0.0871, 0.0910, 0.0948, 0.0987, 0.1026, 0.1064,
0.1103, 0.1141, 0.1179, 0.1217, 0.1255, 0.1293, 0.1331, 0.1368, 0.1406, 0.1443,
0.1480, 0.1517, 0.1554, 0.1591, 0.1628, 0.1664, 0.1700, 0.1736, 0.1772, 0.1808,
0.1844, 0.1879, 0.1915, 0.1950, 0.1985, 0.2019, 0.2054, 0.2088, 0.2123, 0.2157,
0.2190, 0.2224, 0.2257, 0.2291, 0.2324, 0.2357, 0.2389, 0.2422, 0.2454, 0.2486,
0.2517, 0.2549, 0.2580, 0.2611, 0.2642, 0.2673, 0.2704, 0.2734, 0.2764, 0.2794,
0.2823, 0.2852, 0.2881, 0.2910, 0.2939, 0.2967, 0.2995, 0.3023, 0.3051, 0.3078,
0.3106, 0.3133, 0.3159, 0.3186, 0.3212, 0.3238, 0.3264, 0.3289, 0.3315, 0.3340,
0.3365, 0.3389, 0.3413, 0.3438, 0.3461, 0.3485, 0.3508, 0.3531, 0.3554, 0.3577,
0.3599, 0.3621, 0.3643, 0.3665, 0.3686, 0.3708, 0.3729, 0.3749, 0.3770, 0.3790,
0.3810, 0.3830, 0.3849, 0.3869, 0.3888, 0.3907, 0.3925, 0.3944, 0.3962, 0.3980,
0.3997, 0.4015, 0.4032, 0.4049, 0.4066, 0.4082, 0.4099, 0.4115, 0.4131, 0.4147,
0.4162, 0.4177, 0.4192, 0.4207, 0.4222, 0.4236, 0.4251, 0.4265, 0.4279, 0.4292,
0.4306, 0.4319, 0.4332, 0.4345, 0.4357, 0.4370, 0.4382, 0.4394, 0.4406, 0.4418,
0.4429, 0.4441, 0.4452, 0.4463, 0.4474, 0.4484, 0.4495, 0.4505, 0.4515, 0.4525,
0.4535, 0.4545, 0.4554, 0.4564, 0.4573, 0.4582, 0.4591, 0.4599, 0.4608, 0.4616,
0.4625, 0.4633, 0.4641, 0.4649, 0.4656, 0.4664, 0.4671, 0.4678, 0.4686, 0.4693,
0.4699, 0.4706, 0.4713, 0.4719, 0.4726, 0.4732, 0.4738, 0.4744, 0.4750, 0.4756,
0.4761, 0.4767, 0.4772, 0.4778, 0.4783, 0.4788, 0.4793, 0.4798, 0.4803, 0.4808,
0.4812, 0.4817, 0.4821, 0.4826, 0.4830, 0.4834, 0.4838, 0.4842, 0.4846, 0.4850,
0.4854, 0.4857, 0.4861, 0.4864, 0.4868, 0.4871, 0.4875, 0.4878, 0.4881, 0.4884,
0.4887, 0.4890, 0.4893, 0.4896, 0.4898, 0.4901, 0.4904, 0.4906, 0.4909, 0.4911,
0.4913, 0.4916, 0.4918, 0.4920, 0.4922, 0.4925, 0.4927, 0.4929, 0.4931, 0.4932,
0.4934, 0.4936, 0.4938, 0.4940, 0.4941, 0.4943, 0.4945, 0.4946, 0.4948, 0.4949,
0.4951, 0.4952, 0.4953, 0.4955, 0.4956, 0.4957, 0.4959, 0.4960, 0.4961, 0.4962,
0.4963, 0.4964, 0.4965, 0.4966, 0.4967, 0.4968, 0.4969, 0.4970, 0.4971, 0.4972,
0.4973, 0.4974, 0.4974, 0.4975, 0.4976, 0.4977, 0.4977, 0.4978, 0.4979, 0.4979,
0.4980, 0.4981, 0.4981, 0.4982, 0.4982, 0.4983, 0.4984, 0.4984, 0.4985, 0.4985,
0.4986, 0.4986, 0.4987, 0.4987, 0.4987, 0.4988, 0.4988, 0.4989, 0.4989, 0.4989,
0.4990, 0.4990]

def sd_to_area(sd):
  sign = 1
  if sd < 0:
```

```
      sign = -1
   sd = math.fabs(sd)  # get the absolute value of sd
   index = int(sd * 100)
   if len(area_to_sd_map) <= index:
      return sign * area_to_sd_map[-1] # return last element in array
   if index == (sd * 100):
      return sign * area_to_sd_map[index]
   return sign * (area_to_sd_map[index] + area_to_sd_map[index + 1]) / 2


def area_to_sd(area):
   sign = 1
   if area < 0:
      sign = -1
   area = math.fabs(area)
   for a in range(len(area_to_sd_map)):
      if area == area_to_sd_map[a]:
         return sign * a / 100
      if 0 < a and area_to_sd_map[a - 1] < area and area < area_to_sd_map[a]:
         # our area is between this value and the previous
         # for simplicity, we will just take the sd half way between a - 1 and a
         return sign * (a - .5) / 100
   return sign * (len(area_to_sd_map) - 1) / 100


def bootstrap(x):
        samp_x = []
        for i in range(len(x)):
                samp_x.append(random.choice(x))
        return samp_x


def sumofsq(vals, mean):
   # the sum of squares for a group is calculated
   # by getting the sum of the squared difference
   # of each value and the mean of the group that value belongs to
   count = len(vals)
   total = 0
   for i in range (count):
      diff_sq = (vals[i] - mean)**2
      total += diff_sq
   return total


def weightedsumofsq(vals, weights, mean):
   count = len(vals)
   total = 0
   for i in range (count):
      diff_sq = (vals[i] - mean)**2
      total += weights[i] * diff_sq
   return total


# expects list of lists
def onewayanova(grps):
   num_grps = len(grps)
   within_group_means = []
```

```
grp_counts = []
within_ss = 0
total_sum = 0
total_count = 0
for i in range (num_grps):
   grp = grps[i]
   grp_count = len(grp)
   grp_sum = sum(grp)
   within_group_mean = grp_sum / float(grp_count)

   grp_counts.append(grp_count)

   total_count += grp_count# add to total number of vals
   total_sum += grp_sum

   within_group_means.append(within_group_mean)

   # to get the within group sum of squares:
   # sum the following: for every element in the overall group
   # subtract that element's value from that element's group mean
   # square the difference
   # get within group sum of squares
   # this is calculated by summing each group's sum of squares
   within_ss += sumofsq(grp, within_group_mean)

total_mean = total_sum / total_count

# to get the between group sum of squares:
# sum the following: for every element in the overall group
# subtract that element's group mean from the overall group mean
# square the difference
# grp_counts are used as weights
between_ss = weightedsumofsq(within_group_means, grp_counts, total_mean)

# now we want to find out how different the groups are between each other
# compared to how much the values vary within the groups
# if all groups vary a lot within themselves,
# and there is no significant difference
# between the groups, then we expect the differences between
# the groups to vary by about the same amount
# so lets get the ratio of the between group variance and
# the within group variance
# if the ratio is 1, then there is no difference between the groups
# if it is significantly larger than one,
# then there is a significant difference between the groups
# remember: even if the groups are significantly different,
# we still won't know which groups are different

# the between group degrees of freedom
# is equal to the number of groups - 1
# this is because once we know the number of groups - 1,
# we know the last group
```

```
    between_df = len(grp_counts) - 1

    # the within group degress of freedom
    # is equal to the total number of values minus the number of groups
    # this is because for each group, once we know the count - 1 values,
    # we know the last value for that group
    # so we lose the number of groups * 1 degrees of freedom
    within_df = total_count - num_grps

    within_var = within_ss / within_df
    between_var = between_ss / between_df

    f_stat = between_var / within_var

    return f_stat


####################################
#
# Computations
#
####################################

# list of lists
samples = []

# file must be in FASTA format
infile=open(input_file)
for line in infile:
        if line.startswith('>'):
                # start of new sample
                samples.append([])
        elif not line.isspace():
                # line must contain values for previous sample
                samples[len(samples) - 1] += map(float,line.split())
infile.close()

observed_f_statistic = onewayanova(samples)

num_resamples = 10000   # number of times we will resample from our original samples
num_below_observed = 0   # count the number of bootstrap values below the observed
sample statistic
out = []                # will store results of each time we resample

for i in range(num_resamples):
  # get bootstrap samples for each of our groups
  # then compute our statistic of interest
  # append statistic to out
  bootstrap_samples = []  # list of lists
  for sample in samples:
    bootstrap_samples.append(bootstrap(sample))
  # now we have a list of new samples, run onewayanova
  boot_f_statistic = onewayanova(bootstrap_samples)
```

```
    if boot_f_statistic < observed_f_statistic:
      num_below_observed += 1
   out.append(boot_f_statistic)

out.sort()

# standard confidence interval computations
conf_interval = 0.9
tails = (1 - conf_interval) / 2

# in case our lower and upper bounds are not integers,
# we decrease the range (the values we include in our interval),
# so that we can keep the same level of confidence
lower_bound = int(math.ceil(num_resamples * tails))
upper_bound = int(math.floor(num_resamples * (1 - tails)))

# bias-corrected confidence interval computations
p = num_below_observed / float(num_resamples)# proportion of bootstrap values below
the observed value

dist_from_center = p - .5# if this is negative, the original is below the center, if
positive, it is above
z_0 = area_to_sd(dist_from_center)

# now we want to find the proportion that should be between the mean and one of the
tails
tail_sds = area_to_sd(conf_interval / 2)
z_alpha_over_2 = 0 - tail_sds
z_1_minus_alpha_over_2 = tail_sds

# in case our lower and upper bounds are not integers,
# we decrease the range (the values we include in our interval),
# so that we can keep the same level of confidence
bias_corr_lower_bound = int(math.ceil(num_resamples * (0.5 +
sd_to_area(z_alpha_over_2 + (2 * z_0)))))
bias_corr_upper_bound =  int(math.floor(num_resamples * (0.5 +
sd_to_area(z_1_minus_alpha_over_2 + (2 * z_0)))))

####################################
#
# Output
#
####################################

# print observed value and then confidence interval
print "Observed F-statistic: %.2f" % observed_f_statistic
print "We have", conf_interval * 100, "% confidence that the true F-statistic",
print "is between: %.2f" % out[int(lower_bound)], "and %.2f" % out[int(upper_bound)]

print "***** Bias Corrected Confidence Interval *****"
print "We have", conf_interval * 100, "% confidence that the true F-statistic",
```

```
print "is between: %.2f" % out[bias_corr_lower_bound], "and %.2f" %
out[bias_corr_upper_bound]
```

TwoWayAnovaSig.py

```
#!/usr/bin/python

###################################
# Two-Way ANOVA Significance Test
# From: Statistics is Easy! By Dennis Shasha and Manda Wilson
#
#
# Assuming that there is no interaction between the three drugs used and sex, tests to
see the probability
# of getting a f-statistic by chance alone greater than or equal to the one observed.
# Uses shuffling to get a distribution to compare
# to the observed f-statistic.
#
# Author: Manda Wilson
#
# Example of FASTA formatted input file:
# >grp 1 1
# 15 12 13 16
# >grp 2 1
# 19 17 16 15
# >grp 3 1
# 14 13 12 17
# >grp 1 2
# 13 13 12 11
# >grp 2 2
# 13 11 11 17
# >grp 3 2
# 11 12 10
#
# Note that the input file is a little different from our other input files.
# On each descriptive line (beginning with '>') there is the group name (which is
ignored)
# followed by a whitespace character, followed by a number, then another whitespace
character,
# then another number.  These numbers represent indexes in a two-dimensional matrix.
So ">grp 1 1"
# represents the element (a list) at row 1, column 1, ">grp 2 1" represents the
element (a list)
# at row 2, column 1, etc.  This is how we seperate the groups by two factors.  The
rows are
# different categories within one factor, the columns different categories within
another factor.
#
# Pseudocode:
#
# 1. Calculate f-statistic for the observed values (in our example it is 0.93).
#     a. Initialize within sum of squares (wss), total sum (ts), and
#         total count (tc) to 0
```

```
#     b. For each row: (here we loop through the categories of factor a)
#        For each column: (here we loop through the categories of factor b)
#           i. Within group mean (wgm) = sum of group values / number of values in group
#        ii. Store the number of values in group
#        iii. Sum the following: for each value in the group
#              I. Subtract the value from wgm
#              II. Square the result of step (1biiiI)
#              III.  Add the result of step (1biiiII) to wss
#     c. Total mean (tm) = ts / tc
#     d. Mean group means (mgm) = sum(wgm) / num_groups
#     e. Total sum of squares (tss) =
#     i. Sum the following: for each value (include all values)
#              I. Subtract the value from tm
#              II. Square the result of step (1eiI)
#     III.   Add the result of step (1eiII) to tss
#     d. Factor a sum of squares (fass) =
#        i. Sum the following: for each category in factor a
#              I. Subtract the sum of values in this category from tm
#              II. Square the result of step (1diI)
#              III. Multiply the result of step (1diII) by
#                   the number of values in that category
#     e. Factor b sum of squares (fbss) =
#        i. Sum the following: for each category in factor b
#              I. Subtract the sum of values in this category from tm
#              II. Square the result of step (1eiI)
#              III. Multiply the result of step (1eiII) by
#                   the number of values in that category
#     f. Between group sum of squares (bss) = tss - wss
#     g. Factor sum of squares (fss) = bss - (fass + fbss)
#     h. Between degrees of freedom (bdf) = number of groups - 1
#     i. Within degrees of freedom (wdf) = tc - number of groups
#     j. Factor a degrees of freedom (fadf) = number of categories in factor_a - 1
#     k. Factor b degrees of freedom (fbdf) = number of categories in factor_b - 1
#     l. Interaction degrees of freedom (idf) = bdf - fadf - fbdf
#     m. Within group variance (wgv) = wss / wdf
#     n. Between group variance (bgv) = bss / bdf
#     o. Interaction variance (iv) = fss / idf
#     i. f-statistic = iv / wgv
#
# 2. Set a counter to 0, this will count the number of times we get a f-statistic
#     greater than or equal to our observed f-statistic (in our example 0.93).
#
# 3. Do the following 10,000 times:
#     a. Shuffle the observed values. To do this:
#        i. Put the values from all the groups into one array
#        ii. Shuffle the pooled values
#        iii. Reassign the pooled values to groups of the same size as the original
groups
#     b. Calculate f-statistic on the results from step (3a), just as we did in step (1)
#     c. If the result from step (3b) is greater than or equal to our observed f-
statistic (0.93),
#        increment our counter from step (2).
```

```
#
# 3. counter / 10,000 equals the probability of getting a f-statistic greater than or
equal to
#      our observed f-stat (0.93), assuming there is no difference between the groups
#
####################################

import random

####################################
#
# Adjustable variables
#
####################################

input_file = 'input/TwoWayAnova.vals'

####################################
#
# Subroutines
#
####################################

# a list of lists
def shuffle(grps):
   num_rows = len(grps)
   num_cols = 0
   pool = []

   # throw all values together
   for r in range(num_rows):
      num_cols = len(grps[r])
      for c in range(num_cols):
         pool.extend(grps[r][c])
   # mix them up
   random.shuffle(pool)
   # reassign to groups
   new_grps = []
   start_index = 0
   end_index = 0
   for r in range(num_rows):
      new_grps.append([]);
      for c in range(num_cols):
         new_grps[r].append([])
         end_index = start_index + len(grps[r][c])
         new_grps[r][c] = pool[start_index:end_index]
         start_index = end_index
   return new_grps

def sumofsq(vals, mean):
   # the sum of squares for a group is calculated
   # by getting the sum of the squared difference
```

```
  # of each value and the mean of the group that value belongs to
  count = len(vals)
  total = 0
  for i in range (count):
     diff_sq = (vals[i] - mean)**2
     total += diff_sq
  return total

def weightedsumofsq(vals, weights, mean):
  count = len(vals)
  total = 0
  for i in range (count):
     diff_sq = (vals[i] - mean)**2
     total += weights[i] * diff_sq
  return total


# expects list of lists
def twowayanova(grps):
  num_rows = len(grps)# equals the number of categories in factor a
  num_cols = len(grps[0])  # equals the number of categories in factor b
  num_grps = num_rows * num_cols
  within_group_means = []
  grp_counts = []
  within_ss = 0
  total_sum = 0
  total_count = 0
  factor_a = []  # will have num_rows categories, each cateory has num_cols lists
added to it
  factor_b = []  # will have num_cols categories, each cateory has num_rows lists
added to it
  all_vals = []

  for r in range (num_rows):# looping through factor a categories
     if len(factor_a) <= r:# if this category was not initialized yet, initialize it
        factor_a.append([])
     for c in range (num_cols):    # looping through factor b categories
        if len(factor_b) <= c:   # if this category was not initialized yet, initialize
it
           factor_b.append([])
        grp = grps[r][c]
        all_vals += grp
        factor_a[r] += grp
        factor_b[c] += grp
        grp_count = float(len(grp))
        grp_sum = sum(grp)
        within_group_mean = grp_sum / float(grp_count)

        grp_counts.append(grp_count)

        total_count += grp_count# add to total number of vals
        total_sum += grp_sum
```

```python
        within_group_means.append(within_group_mean)

        # to get the within group sum of squares:
        # sum the following: for every element in the overall group
        # subtract that element's value from that element's group mean
        # square the difference
        # get within group sum of squares
        # this is calculated by summing each group's sum of squares
        within_ss += sumofsq(grp, within_group_mean)

    total_mean = total_sum / float(total_count)
    mean_grp_means = sum(within_group_means) / float(num_grps)

    # total sum of squares
    total_ss = sumofsq(all_vals, total_mean)

    factor_a_ss = weightedsumofsq(map(lambda x: sum(x) / float(len(x)), factor_a),
map(lambda x: float(len(x)), factor_a), total_mean)
    factor_b_ss = weightedsumofsq(map(lambda x: sum(x) / float(len(x)), factor_b),
map(lambda x: float(len(x)), factor_b), total_mean)

    # get between group sum of squares
    # NOTE: to compute the between group sum of squares:
    # sum the following: for every element in the overall group
    # subtract that element's group mean from the overall group mean
    # square the difference
    between_ss = total_ss - within_ss
    # NOTE: we could have done the following
    # between_ss = weightedsumofsq(within_group_means, grp_counts, total_mean)

    factor_ss = between_ss - (factor_a_ss + factor_b_ss)

    # now we want to find out how different the groups are between each other
    # compared to how much the values vary within the groups
    # if all groups vary a lot within themselves, and there is no significant difference
    # between the groups, then we expect the differences between the groups to vary by
about the same amount
    # so lets get the ratio of the between group variance and the within group variance
    # if the ratio is 1, then there is no difference between the groups
    # if it is significantly larger than one, then there is a significant difference
between the groups
    # remember: even if the groups are significantly different, we still won't know
which groups are different

    # the between group degrees of freedom
    # is equal to the number of groups - 1
    # this is because once we know the number of groups - 1, we know the last group
    between_df = num_grps - 1

    # the within group degress of freedom
    # is equal to the total number of values minus the number of groups
```

```
    # this is because for each group, once we know the count - 1 values, we know the
last value for that group
    # so we lose the number of groups * 1 degrees of freedom
    within_df = total_count - num_grps

    # the interaction degrees of freedom
    # is equal to between group degrees of freedom
    # minus factor a degrees of freedom
    # minus factor b degrees of freedom
    factor_a_df = num_rows - 1     # number of categories in factor_a - 1
    factor_b_df = num_cols - 1     # number of categories in factor_a - 1
    interaction_df = between_df - factor_a_df - factor_b_df

    within_var = within_ss / within_df
    between_var = between_ss / between_df
    interaction_var = factor_ss / interaction_df

    # return our f-statistic
    return interaction_var / within_var


#####################################
#
# Computations
#
#####################################


# list of lists
samples = []

# file must be in FASTA format
infile=open(input_file)
r = 0
c = 0
for line in infile:
    if line.startswith('>'):
        # read in index
        tokens = line.split()     # split on whitespace
        c = int(tokens[-1]) - 1  # column index is last token on line, we subtract one b/
c python indexes start at 0
        r = int(tokens[-2]) - 1  # row index is second to last token on line, we subtract
one b/c python indexes start at 0
        # make sure we have enough rows in our samples matrix
        while r >= len(samples):
            samples.append([])
        # now make sure that row has enough columns
        while c >= len(samples[r]):
            samples[r].append([])
    elif not line.isspace():
        # line must contain values for previous sample
        samples[r][c] += map(float,line.split())
infile.close()
```

```
observed_f_statistic = twowayanova(samples)

count = 0
num_shuffles = 10000

for i in range(num_shuffles):
  new_samples = shuffle(samples)
  f_statistic = twowayanova(new_samples)
  if (f_statistic >= observed_f_statistic):
    count = count + 1



###################################
#
# Output
#
###################################

print "Observed F-statistic: %.2f" % observed_f_statistic
print count, "out of 10000 experiments had a F-statistic greater than or equal to
%.2f" % observed_f_statistic
print "Probability that chance alone gave us a F-statistic",
print "of %.2f or more" % observed_f_statistic, "is", (count / float(num_shuffles))
```

TwoWayAnova.vals

```
>grp 1 1
15 12 13 16
>grp 2 1
19 17 16 15
>grp 3 1
14 13 12 17
>grp 1 2
13 13 12 11
>grp 2 2
13 11 11 17
>grp 3 2
11 12 10
```

TwoWayAnovaConf.py

```
#!/usr/bin/python

###################################
# Two-Way Confidence Interval
# From: Statistics is Easy! By Dennis Shasha and Manda Wilson
#
#
# Uses shuffling & bootstrapping to get a 90% confidence interval for the f-statistic.
#
# Author: Manda Wilson
#
# Example of FASTA formatted input file:
```

```
# >grp 1 1
# 15 12 13 16
# >grp 2 1
# 19 17 16 15
# >grp 3 1
# 14 13 12 17
# >grp 1 2
# 13 13 12 11
# >grp 2 2
# 13 11 11 17
# >grp 3 2
# 11 12 10
#
# Included in the code, but NOT in the pseudocode,
# is the bias-corrected confidence interval.
# See http://www.statisticsiseasy.com/BiasCorrectedConfInter.html
# for more information on bias-corrected cofidence intervals.
#
# Note that the input file is a little different from our other input files.
# On each descriptive line (beginning with '>') there is the group name (which is
ignored)
# followed by a whitespace character, followed by a number, then another whitespace
character,
# then another number.  These numbers represent indexes in a two-dimensional matrix.
So ">grp 1 1"
# represents the element (a list) at row 1, column 1, ">grp 2 1" represents the
element (a list)
# at row 2, column 1, etc.  This is how we seperate the groups by two factors.  The
rows are
# different categories within one factor, the columns different categories within
another factor.
#
# Pseudocode:
#
# 1. Calculate f-statistic for the observed values (in our example it is 0.93).
#     a. Initialize within sum of squares (wss), total sum (ts), and
#        total count (tc) to 0
#     b. For each row: (here we loop through the categories of factor a)
#        For each column: (here we loop through the categories of factor b)
#        i. Within group mean (wgm) = sum of group values / number of values in group
#     ii. Store the number of values in group
#     iii. Sum the following: for each value in the group
#             I. Subtract the value from wgm
#             II. Square the result of step (1biiiI)
#             III.  Add the result of step (1biiiII) to wss
#     c. Total mean (tm) = ts / tc
#     d. Mean group means (mgm) = sum(wgm) / num_groups
#     e. Total sum of squares (tss) =
#     i. Sum the following: for each value (include all values)
#             I. Subtract the value from tm
#             II. Square the result of step (1eiI)
#        III.  Add the result of step (1eiII) to tss
```

```
#    d. Factor a sum of squares (fass) =
#        i. Sum the following: for each category in factor a
#            I. Subtract the sum of values in this category from tm
#            II. Square the result of step (1diI)
#            III. Multiply the result of step (1diII) by
#                the number of values in that category
#    e. Factor b sum of squares (fbss) =
#        i. Sum the following: for each category in factor b
#            I. Subtract the sum of values in this category from tm
#            II. Square the result of step (1eiI)
#            III. Multiply the result of step (1eiII) by
#                the number of values in that category
#    f. Between group sum of squares (bss) = tss - wss
#    g. Factor sum of squares (fss) = bss - (fass + fbss)
#    h. Between degrees of freedom (bdf) = number of groups - 1
#    i. Within degrees of freedom (wdf) = tc - number of groups
#    j. Factor a degrees of freedom (fadf) = number of categories in factor_a - 1
#    k. Factor b degrees of freedom (fbdf) = number of categories in factor_b - 1
#    l. Interaction degrees of freedom (idf) = bdf - fadf - fbdf
#    m. Within group variance (wgv) = wss / wdf
#    n. Between group variance (bgv) = bss / bdf
#    o. Interaction variance (iv) = fss / idf
#    i. f-statistic = iv / wgv
#
# 2. Do the following 10,000 times:
#    a. For each sample we have get a bootstrap sample:
#        i. Create a new array of the same size as the original sample
#        ii. Fill the array with randomly picked values from the original sample
(randomly picked with replacement)
#    b. Calculate the f-statistic on the bootstrap samples, just as we did in step (1)
#        with the original samples.
#
# 3. Sort the f-statistics computed in step (2).
#
# 4. Compute the size of each interval tail.  If we want a 90% confidence interval,
then 1 - 0.9 yields the
#    portion of the interval in the tails.  We divide this by 2 to get the size of each
tail, in this case 0.05.
#
# 5. Compute the upper and lower bounds.  To get the lower bound we multiply the tail
size by the number of
#    bootstraps we ran and round this value up to the nearest integer (to make sure
this value maps
#    to an actual boostrap we ran).  In this case we multiple 0.05 by 10,000 and get
500, which rounds up to 500.
#
#    To compute the upper bound we subtract the tail size from 1 and then multiply that
by the number of bootstraps
#    we ran.  Round the result of the last step down to the nearest integer.  Again,
this is to ensure this value
#    maps to an actual bootstrap we ran.  We round the lower bound up and the upper
bound down to reduce the
```

```
#    confidence interval size, so that we can still say we have as much confidence in
this result.
#
# 6. The bootstrap values at the lower bound and upper bound give us our confidence
interval.
#
####################################

import random
import math
import sys


####################################
#
# Adjustable variables
#
####################################

input_file = 'input/TwoWayAnova.vals'


####################################
#
# Subroutines
#
####################################

# maps proportion of values above statistic
# to number of standard deviations above statistic
# keys will be index / 100 \:[0-9]\.[0-9][0-9]\,
area_to_sd_map = [0.0000, 0.0040, 0.0080, 0.0120, 0.0160, 0.0199, 0.0239, 0.0279,
0.0319, 0.0359, 0.0398, 0.0438, 0.0478, 0.0517, 0.0557, 0.0596, 0.0636, 0.0675,
0.0714, 0.0753, 0.0793, 0.0832, 0.0871, 0.0910, 0.0948, 0.0987, 0.1026, 0.1064,
0.1103, 0.1141, 0.1179, 0.1217, 0.1255, 0.1293, 0.1331, 0.1368, 0.1406, 0.1443,
0.1480, 0.1517, 0.1554, 0.1591, 0.1628, 0.1664, 0.1700, 0.1736, 0.1772, 0.1808,
0.1844, 0.1879, 0.1915, 0.1950, 0.1985, 0.2019, 0.2054, 0.2088, 0.2123, 0.2157,
0.2190, 0.2224, 0.2257, 0.2291, 0.2324, 0.2357, 0.2389, 0.2422, 0.2454, 0.2486,
0.2517, 0.2549, 0.2580, 0.2611, 0.2642, 0.2673, 0.2704, 0.2734, 0.2764, 0.2794,
0.2823, 0.2852, 0.2881, 0.2910, 0.2939, 0.2967, 0.2995, 0.3023, 0.3051, 0.3078,
0.3106, 0.3133, 0.3159, 0.3186, 0.3212, 0.3238, 0.3264, 0.3289, 0.3315, 0.3340,
0.3365, 0.3389, 0.3413, 0.3438, 0.3461, 0.3485, 0.3508, 0.3531, 0.3554, 0.3577,
0.3599, 0.3621, 0.3643, 0.3665, 0.3686, 0.3708, 0.3729, 0.3749, 0.3770, 0.3790,
0.3810, 0.3830, 0.3849, 0.3869, 0.3888, 0.3907, 0.3925, 0.3944, 0.3962, 0.3980,
0.3997, 0.4015, 0.4032, 0.4049, 0.4066, 0.4082, 0.4099, 0.4115, 0.4131, 0.4147,
0.4162, 0.4177, 0.4192, 0.4207, 0.4222, 0.4236, 0.4251, 0.4265, 0.4279, 0.4292,
0.4306, 0.4319, 0.4332, 0.4345, 0.4357, 0.4370, 0.4382, 0.4394, 0.4406, 0.4418,
0.4429, 0.4441, 0.4452, 0.4463, 0.4474, 0.4484, 0.4495, 0.4505, 0.4515, 0.4525,
0.4535, 0.4545, 0.4554, 0.4564, 0.4573, 0.4582, 0.4591, 0.4599, 0.4608, 0.4616,
0.4625, 0.4633, 0.4641, 0.4649, 0.4656, 0.4664, 0.4671, 0.4678, 0.4686, 0.4693,
0.4699, 0.4706, 0.4713, 0.4719, 0.4726, 0.4732, 0.4738, 0.4744, 0.4750, 0.4756,
0.4761, 0.4767, 0.4772, 0.4778, 0.4783, 0.4788, 0.4793, 0.4798, 0.4803, 0.4808,
0.4812, 0.4817, 0.4821, 0.4826, 0.4830, 0.4834, 0.4838, 0.4842, 0.4846, 0.4850,
0.4854, 0.4857, 0.4861, 0.4864, 0.4868, 0.4871, 0.4875, 0.4878, 0.4881, 0.4884,
0.4887, 0.4890, 0.4893, 0.4896, 0.4898, 0.4901, 0.4904, 0.4906, 0.4909, 0.4911,
0.4913, 0.4916, 0.4918, 0.4920, 0.4922, 0.4925, 0.4927, 0.4929, 0.4931, 0.4932,
```

```
0.4934, 0.4936, 0.4938, 0.4940, 0.4941, 0.4943, 0.4945, 0.4946, 0.4948, 0.4949,
0.4951, 0.4952, 0.4953, 0.4955, 0.4956, 0.4957, 0.4959, 0.4960, 0.4961, 0.4962,
0.4963, 0.4964, 0.4965, 0.4966, 0.4967, 0.4968, 0.4969, 0.4970, 0.4971, 0.4972,
0.4973, 0.4974, 0.4974, 0.4975, 0.4976, 0.4977, 0.4977, 0.4978, 0.4979, 0.4979,
0.4980, 0.4981, 0.4981, 0.4982, 0.4982, 0.4983, 0.4984, 0.4984, 0.4985, 0.4985,
0.4986, 0.4986, 0.4987, 0.4987, 0.4987, 0.4988, 0.4988, 0.4989, 0.4989, 0.4989,
0.4990, 0.4990]

def sd_to_area(sd):
    sign = 1
    if sd < 0:
        sign = -1
    sd = math.fabs(sd)  # get the absolute value of sd
    index = int(sd * 100)
    if len(area_to_sd_map) <= index:
        return sign * area_to_sd_map[-1] # return last element in array
    if index == (sd * 100):
        return sign * area_to_sd_map[index]
    return sign * (area_to_sd_map[index] + area_to_sd_map[index + 1]) / 2

def area_to_sd(area):
    sign = 1
    if area < 0:
        sign = -1
    area = math.fabs(area)
    for a in range(len(area_to_sd_map)):
        if area == area_to_sd_map[a]:
            return sign * a / 100
        if 0 < a and area_to_sd_map[a - 1] < area and area < area_to_sd_map[a]:
            # our area is between this value and the previous
            # for simplicity, we will just take the sd half way between a - 1 and a
            return sign * (a - .5) / 100
    return sign * (len(area_to_sd_map) - 1) / 100

def bootstrap(x):
        samp_x = []
        for i in range(len(x)):
                samp_x.append(random.choice(x))
        return samp_x

def sumofsq(vals, mean):
    # the sum of squares for a group is calculated
    # by getting the sum of the squared difference
    # of each value and the mean of the group that value belongs to
    count = len(vals)
    total = 0
    for i in range (count):
        diff_sq = (vals[i] - mean)**2
        total += diff_sq
    return total

def weightedsumofsq(vals, weights, mean):
```

```
  count = len(vals)
  total = 0
  for i in range (count):
     diff_sq = (vals[i] - mean)**2
     total += weights[i] * diff_sq
  return total


# expects list of lists
def twowayanova(grps):
  num_rows = len(grps)# equals the number of categories in factor a
  num_cols = len(grps[0])  # equals the number of categories in factor b
  num_grps = num_rows * num_cols
  within_group_means = []
  grp_counts = []
  within_ss = 0
  total_sum = 0
  total_count = 0
  factor_a = []  # will have num_rows categories, each cateory has num_cols lists
added to it
  factor_b = []  # will have num_cols categories, each cateory has num_rows lists
added to it
  all_vals = []

  for r in range (num_rows):# looping through factor a categories
     if len(factor_a) <= r:# if this category was not initialized yet, initialize it
        factor_a.append([])
     for c in range (num_cols):    # looping through factor b categories
        if len(factor_b) <= c:   # if this category was not initialized yet, initialize
it
           factor_b.append([])
        grp = grps[r][c]
        all_vals += grp
        factor_a[r] += grp
        factor_b[c] += grp
        grp_count = float(len(grp))
        grp_sum = sum(grp)
        within_group_mean = grp_sum / float(grp_count)

        grp_counts.append(grp_count)

        total_count += grp_count# add to total number of vals
        total_sum += grp_sum

        within_group_means.append(within_group_mean)

        # to get the within group sum of squares:
        # sum the following: for every element in the overall group
        # subtract that element's value from that element's group mean
        # square the difference
        # get within group sum of squares
        # this is calculated by summing each group's sum of squares
        within_ss += sumofsq(grp, within_group_mean)
```

```
    total_mean = total_sum / float(total_count)
    mean_grp_means = sum(within_group_means) / float(num_grps)


    # total sum of squares
    total_ss = sumofsq(all_vals, total_mean)


    factor_a_ss = weightedsumofsq(map(lambda x: sum(x) / float(len(x)), factor_a),
map(lambda x: float(len(x)), factor_a), total_mean)
    factor_b_ss = weightedsumofsq(map(lambda x: sum(x) / float(len(x)), factor_b),
map(lambda x: float(len(x)), factor_b), total_mean)


    # get between group sum of squares
    # NOTE: to compute the between group sum of squares:
    # sum the following: for every element in the overall group
    # subtract that element's group mean from the overall group mean
    # square the difference
    between_ss = total_ss - within_ss
    # NOTE: we could have done the following
    # between_ss = weightedsumofsq(within_group_means, grp_counts, total_mean)


    factor_ss = between_ss - (factor_a_ss + factor_b_ss)


    # now we want to find out how different the groups are between each other
    # compared to how much the values vary within the groups
    # if all groups vary a lot within themselves, and there is no significant difference
    # between the groups, then we expect the differences between the groups to vary by
about the same amount
    # so lets get the ratio of the between group variance and the within group variance
    # if the ratio is 1, then there is no difference between the groups
    # if it is significantly larger than one, then there is a significant difference
between the groups
    # remember: even if the groups are significantly different, we still won't know
which groups are different


    # the between group degrees of freedom
    # is equal to the number of groups - 1
    # this is because once we know the number of groups - 1, we know the last group
    between_df = num_grps - 1


    # the within group degress of freedom
    # is equal to the total number of values minus the number of groups
    # this is because for each group, once we know the count - 1 values, we know the
last value for that group
    # so we lose the number of groups * 1 degrees of freedom
    within_df = total_count - num_grps


    # the interaction degrees of freedom
    # is equal to between group degrees of freedom
    # minus factor a degrees of freedom
    # minus factor b degrees of freedom
    factor_a_df = num_rows - 1    # number of categories in factor_a - 1
```

```
    factor_b_df = num_cols - 1    # number of categories in factor_a - 1
    interaction_df = between_df - factor_a_df - factor_b_df

    within_var = within_ss / within_df
    between_var = between_ss / between_df
    interaction_var = factor_ss / interaction_df

    # return our f-statistic
    return interaction_var / within_var


###################################
#
# Computations
#
###################################


# list of lists
samples = []

# file must be in FASTA format
infile=open(input_file)
r = 0
c = 0
for line in infile:
   if line.startswith('>'):
      # read in index
      tokens = line.split()    # split on whitespace
      c = int(tokens[-1]) - 1  # column index is last token on line, we subtract one b/
c python indexes start at 0
      r = int(tokens[-2]) - 1  # row index is second to last token on line, we subtract
one b/c python indexes start at 0
      # make sure we have enough rows in our samples matrix
      while r >= len(samples):
         samples.append([])
      # now make sure that row has enough columns
      while c >= len(samples[r]):
         samples[r].append([])
   elif not line.isspace():
      # line must contain values for previous sample
      samples[r][c] += map(float,line.split())
infile.close()

observed_f_statistic = twowayanova(samples)

num_resamples = 10000    # number of times we will resample from our original samples
num_below_observed = 0   # count the number of bootstrap values below the observed
sample statistic
out = []  # will store results of each time we resample

for i in range(num_resamples):
   # get bootstrap samples for each of our groups
   # then compute our statistic of interest
```

```
    # append statistic to out
    bootstrap_samples = []  # list of lists
    r = 0
    c = 0
    for r in range(len(samples)):
      bootstrap_samples.append([])
      for c in range(len(samples[r])):
        bootstrap_samples[r].append(bootstrap(samples[r][c]))
    # now we have a list of new samples, run onewayanova
    boot_f_statistic = twowayanova(bootstrap_samples)
    if boot_f_statistic < observed_f_statistic:
      num_below_observed += 1
    out.append(boot_f_statistic)

out.sort()

# standard confidence interval computations
conf_interval = 0.9
tails = (1 - conf_interval) / 2

# in case our lower and upper bounds are not integers,
# we decrease the range (the values we include in our interval),
# so that we can keep the same level of confidence
lower_bound = int(math.ceil(num_resamples * tails))
upper_bound = int(math.floor(num_resamples * (1 - tails)))

# bias-corrected confidence interval computations
p = num_below_observed / float(num_resamples)# proportion of bootstrap values below
the observed value

dist_from_center = p - .5# if this is negative, the original is below the center, if
positive, it is above
z_0 = area_to_sd(dist_from_center)

# now we want to find the proportion that should be between the mean and one of the
tails
tail_sds = area_to_sd(conf_interval / 2)
z_alpha_over_2 = 0 - tail_sds
z_1_minus_alpha_over_2 = tail_sds

# in case our lower and upper bounds are not integers,
# we decrease the range (the values we include in our interval),
# so that we can keep the same level of confidence
bias_corr_lower_bound = int(math.ceil(num_resamples * (0.5 +
sd_to_area(z_alpha_over_2 + (2 * z_0)))))
bias_corr_upper_bound = int(math.floor(num_resamples * (0.5 +
sd_to_area(z_1_minus_alpha_over_2 + (2 * z_0)))))

####################################
#
# Output
#
```

```
#####################################

# print observed value and then confidence interval
print "Observed F-statistic: %.2f" % observed_f_statistic
print "We have", conf_interval * 100, "% confidence that the true F-statistic",
print "is between: %.2f" % out[int(lower_bound)], "and %.2f" % out[int(upper_bound)]

print "***** Bias Corrected Confidence Interval *****"
print "We have", conf_interval * 100, "% confidence that the true F-statistic",
print "is between: %.2f" % out[bias_corr_lower_bound], "and %.2f" %
out[bias_corr_upper_bound]
```

RegressionSig.py

```
#!/usr/bin/python

#####################################
# Regression Significance Test
# From: Statistics is Easy! By Dennis Shasha and Manda Wilson
#
#
# Assuming that x is not a good predictor of y, tests to see the probability
# of getting a slope by chance alone greater than or equal to the one observed (less
# than or equal to the one observed if the observed slope is negative).
# Uses shuffling to get a distribution to compare
# to the observed f-statistic.
#
# Author: Manda Wilson
#
# Example of FASTA formatted input file:
# >grp_x
# 1350 1510 1420 1210 1250 1300 1580 1310 1290 1320 1490 1200 1360
# >grp_y
# 3.6 3.8 3.7 3.3 3.9 3.4 3.8 3.7 3.5 3.4 3.8 3.0 3.1
#
# Pseudocode:
#
# 1. Calculate the regression line for the observed values (in our example it is y' =
0.0014 x +  1.6333).
#     a. mean_x = sum of x values / number of x values
#     b. mean_y = sum of y values / number of y values
#     c. Calculate the sum of products:
# i. Initialize total to 0
# ii. For each pair of x, y values:
#            I. product = (x - mean_x) * (y - mean_y)
#            II. total += product
#     d. Calculate the sum of squares for the x values:
#        i. Initialize total to 0
#        ii. For each x value:
#            I. diff_sq = (x - mean_x)^2
#            II. total += diff_sq
#     e. b = sum of products / sum of squares for the x values
#     f. a = mean_y - (b * mean_x)
```

```
#     g. Line of best fit is: y' = bx + a
#
# 2. Set a counter to 0, this will count the number of times we get a slope (b)
#     greater than or equal to 0.0014.  Note: if you have a negative slope, count
#     the number of times you get a slope less than or equal to the original negative
slope.
#
# 3. Do the following 10,000 times:
#     a. Shuffle the y values.
#     b. Calculate regression line on the results from step (3a), just as we did in step
(1)
#     c. If the slope (b)  from step (3b) is greater than or equal to our observed slope
(0.0014),
#          increment our counter from step (2).
#
# 3. counter / 10,000 equals the probability of getting a slope greater than
#     or equal to 0.0014, assuming x does not predict y
#
####################################

import random

####################################
#
# Adjustable variables
#
####################################

input_file = 'input/Correlation.vals'

####################################
#
# Subroutines
#
####################################

# a list of lists
def shuffle(grps):
        num_grps = len(grps)
        pool = []

        # throw all values together
        for i in range(num_grps):
                pool.extend(grps[i])
        # mix them up
        random.shuffle(pool)
        # reassign to groups
        new_grps = []
        start_index = 0
        end_index = 0
        for i in range(num_grps):
                end_index = start_index + len(grps[i])
```

```
                    new_grps.append(pool[start_index:end_index])
                    start_index = end_index
        return new_grps

def sumofsq(vals, mean):
   count = len(vals)
   total = 0
   for i in range (count):
      diff_sq = (vals[i] - mean)**2
      total += diff_sq
   return total

def sumofproducts(x_vals, y_vals,  mean_x, mean_y):
   count = len(x_vals)
   total = 0
   for i in range (count):
      product = (x_vals[i] - mean_x) * (y_vals[i] - mean_y)
      total += product
   return total

def regressionline(grp_x, grp_y):
   sum_x = sum(grp_x)
   sum_y = sum(grp_y)

   count_x = float(len(grp_x))
   count_y = float(len(grp_y))

   mean_x = sum_x / count_x
   mean_y = sum_y / count_y

   # get the sum of products
   sum_of_prod = sumofproducts(grp_x, grp_y, mean_x, mean_y)

   # get the sum of squares for x
   sum_of_sq_x = sumofsq(grp_x, mean_x)

   b = sum_of_prod / sum_of_sq_x
   a = mean_y - (b * mean_x)
   return (a, b)


####################################
#
# Computations
#
####################################

# list of lists
samples = []

# file must be in FASTA format
infile=open(input_file)
for line in infile:
```

```
        if line.startswith('>'):
                # start of new sample
                samples.append([])
        elif not line.isspace():
                # line must contain values for previous sample
                samples[len(samples) - 1] += map(float,line.split())
infile.close()

grp_x = samples[0]
grp_y = samples[1]

(observed_a, observed_b) = regressionline(grp_x, grp_y)

count = 0
num_shuffles = 10000

for i in range(num_shuffles):
        new_y_values = shuffle([grp_y])[0]
        (a, b) = regressionline(grp_x, new_y_values)
        if ((observed_b >= 0 and b > observed_b) or (observed_b < 0 and b <
observed_b)):
                count = count + 1


####################################
#
# Output
#
####################################

sign = "+"

if (observed_a < 0):
        sign = "-"

observed_a = abs(observed_a)

print "Line of best fit for observed data: ",
print "y' = %.4f" % observed_b, "x", sign, " %.4f" % observed_a
print count, "out of", num_shuffles, "experiments had a slope",
if observed_b < 0:
        print "less than or equal to",
else:
        print "greater than or equal to",
print "%.4f" % observed_b, "."
print "The chance of getting a slope",
if observed_b < 0:
        print "less than or equal to",
else:
        print "greater than or equal to",
print "%.4f" % observed_b, "is", (count / float(num_shuffles)), "."
```

Correlation.vals

```
>grp_x
1350 1510 1420 1210 1250 1300 1580 1310 1290 1320 1490 1200 1360
>grp_y
3.6 3.8 3.7 3.3 3.9 3.4 3.8 3.7 3.5 3.4 3.8 3.0 3.1
```

RegressionConf.py

```
#!/usr/bin/python

####################################
# Regression Confidence Interval Example
# From: Statistics is Easy! By Dennis Shasha and Manda Wilson
#
#
# Uses shuffling & bootstrapping to get a 90% confidence interval for the slope of the
regression line.
#
# Author: Manda Wilson
#
# Example of FASTA formatted input file:
# >grp_x
# 1350 1510 1420 1210 1250 1300 1580 1310 1290 1320 1490 1200 1360
# >grp_y
# 3.6 3.8 3.7 3.3 3.9 3.4 3.8 3.7 3.5 3.4 3.8 3.0 3.1
#
# Included in the code, but NOT in the pseudocode,
# is the bias-corrected confidence interval.
# See http://www.statisticsiseasy.com/BiasCorrectedConfInter.html
# for more information on bias-corrected cofidence intervals.
#
# Pseudocode:
#
# 1. Calculate the regression line for the observed values (in our example it is y' =
0.0014 x +  1.6333).
#    a. mean_x = sum of x values / number of x values
#    b. mean_y = sum of y values / number of y values
#    c. Calculate the sum of products:
# i. Initialize total to 0
# ii. For each pair of x, y values:
#           I. product = (x - mean_x) * (y - mean_y)
#           II. total += product
#    d. Calculate the sum of squares for the x values:
#       i. Initialize total to 0
#       ii. For each x value:
#           I. diff_sq = (x - mean_x)^2
#           II. total += diff_sq
#    e. b = sum of products / sum of squares for the x values
#    f. a = mean_y - (b * mean_x)
#    g. Line of best fit is: y' = bx + a
#
# 2. Do the following 10,000 times:
```

```
#     a.
#         i. Create two new arrays of the same size as the original sample arrays (one
for x and one for y)
#         ii. Fill the new arrays with randomly picked pairs from the original arrays
(randomly picked with replacement)
#     b. Compute the regression line on the bootstrap samples, just as we did in step
(1)
#        with the original samples.
#
# 3. Sort the differences computed in step (2).
#
# 4. Compute the size of each interval tail.  If we want a 90% confidence interval,
then 1 - 0.9 yields the
#    portion of the interval in the tails.  We divide this by 2 to get the size of each
tail, in this case 0.05.
#
# 5. Compute the upper and lower bounds.  To get the lower bound we multiply the tail
size by the number of
#    bootstraps we ran and round this value up to the nearest integer (to make sure
this value maps
#    to an actual boostrap we ran).  In this case we multiple 0.05 by 10,000 and get
500, which rounds up to 500.
#
#    To compute the upper bound we subtract the tail size from 1 and then multiply that
by the number of bootstraps
#    we ran.  Round the result of the last step down to the nearest integer.  Again,
this is to ensure this value
#    maps to an actual bootstrap we ran.  We round the lower bound up and the upper
bound down to reduce the
#    confidence interval size, so that we can still say we have as much confidence in
this result.
#
# 6. The bootstrap values at the lower bound and upper bound give us our confidence
interval.
#
####################################

import random
import math

####################################
#
# Adjustable variables
#
####################################

input_file = 'input/Correlation.vals'
conf_interval = 0.9

####################################
#
# Subroutines
#
```

```
####################################

# maps proportion of values above statistic
# to number of standard deviations above statistic
# keys will be index / 100 \:[0-9]\.[0-9][0-9]\,
area_to_sd_map = [0.0000, 0.0040, 0.0080, 0.0120, 0.0160, 0.0199, 0.0239, 0.0279,
0.0319, 0.0359, 0.0398, 0.0438, 0.0478, 0.0517, 0.0557, 0.0596, 0.0636, 0.0675,
0.0714, 0.0753, 0.0793, 0.0832, 0.0871, 0.0910, 0.0948, 0.0987, 0.1026, 0.1064,
0.1103, 0.1141, 0.1179, 0.1217, 0.1255, 0.1293, 0.1331, 0.1368, 0.1406, 0.1443,
0.1480, 0.1517, 0.1554, 0.1591, 0.1628, 0.1664, 0.1700, 0.1736, 0.1772, 0.1808,
0.1844, 0.1879, 0.1915, 0.1950, 0.1985, 0.2019, 0.2054, 0.2088, 0.2123, 0.2157,
0.2190, 0.2224, 0.2257, 0.2291, 0.2324, 0.2357, 0.2389, 0.2422, 0.2454, 0.2486,
0.2517, 0.2549, 0.2580, 0.2611, 0.2642, 0.2673, 0.2704, 0.2734, 0.2764, 0.2794,
0.2823, 0.2852, 0.2881, 0.2910, 0.2939, 0.2967, 0.2995, 0.3023, 0.3051, 0.3078,
0.3106, 0.3133, 0.3159, 0.3186, 0.3212, 0.3238, 0.3264, 0.3289, 0.3315, 0.3340,
0.3365, 0.3389, 0.3413, 0.3438, 0.3461, 0.3485, 0.3508, 0.3531, 0.3554, 0.3577,
0.3599, 0.3621, 0.3643, 0.3665, 0.3686, 0.3708, 0.3729, 0.3749, 0.3770, 0.3790,
0.3810, 0.3830, 0.3849, 0.3869, 0.3888, 0.3907, 0.3925, 0.3944, 0.3962, 0.3980,
0.3997, 0.4015, 0.4032, 0.4049, 0.4066, 0.4082, 0.4099, 0.4115, 0.4131, 0.4147,
0.4162, 0.4177, 0.4192, 0.4207, 0.4222, 0.4236, 0.4251, 0.4265, 0.4279, 0.4292,
0.4306, 0.4319, 0.4332, 0.4345, 0.4357, 0.4370, 0.4382, 0.4394, 0.4406, 0.4418,
0.4429, 0.4441, 0.4452, 0.4463, 0.4474, 0.4484, 0.4495, 0.4505, 0.4515, 0.4525,
0.4535, 0.4545, 0.4554, 0.4564, 0.4573, 0.4582, 0.4591, 0.4599, 0.4608, 0.4616,
0.4625, 0.4633, 0.4641, 0.4649, 0.4656, 0.4664, 0.4671, 0.4678, 0.4686, 0.4693,
0.4699, 0.4706, 0.4713, 0.4719, 0.4726, 0.4732, 0.4738, 0.4744, 0.4750, 0.4756,
0.4761, 0.4767, 0.4772, 0.4778, 0.4783, 0.4788, 0.4793, 0.4798, 0.4803, 0.4808,
0.4812, 0.4817, 0.4821, 0.4826, 0.4830, 0.4834, 0.4838, 0.4842, 0.4846, 0.4850,
0.4854, 0.4857, 0.4861, 0.4864, 0.4868, 0.4871, 0.4875, 0.4878, 0.4881, 0.4884,
0.4887, 0.4890, 0.4893, 0.4896, 0.4898, 0.4901, 0.4904, 0.4906, 0.4909, 0.4911,
0.4913, 0.4916, 0.4918, 0.4920, 0.4922, 0.4925, 0.4927, 0.4929, 0.4931, 0.4932,
0.4934, 0.4936, 0.4938, 0.4940, 0.4941, 0.4943, 0.4945, 0.4946, 0.4948, 0.4949,
0.4951, 0.4952, 0.4953, 0.4955, 0.4956, 0.4957, 0.4959, 0.4960, 0.4961, 0.4962,
0.4963, 0.4964, 0.4965, 0.4966, 0.4967, 0.4968, 0.4969, 0.4970, 0.4971, 0.4972,
0.4973, 0.4974, 0.4974, 0.4975, 0.4976, 0.4977, 0.4977, 0.4978, 0.4979, 0.4979,
0.4980, 0.4981, 0.4981, 0.4982, 0.4982, 0.4983, 0.4984, 0.4984, 0.4985, 0.4985,
0.4986, 0.4986, 0.4987, 0.4987, 0.4987, 0.4988, 0.4988, 0.4989, 0.4989, 0.4989,
0.4990, 0.4990]

def sd_to_area(sd):
  sign = 1
  if sd < 0:
    sign = -1
  sd = math.fabs(sd)  # get the absolute value of sd
  index = int(sd * 100)
  if len(area_to_sd_map) <= index:
    return sign * area_to_sd_map[-1] # return last element in array
  if index == (sd * 100):
    return sign * area_to_sd_map[index]
  return sign * (area_to_sd_map[index] + area_to_sd_map[index + 1]) / 2

def area_to_sd(area):
  sign = 1
  if area < 0:
    sign = -1
```

```
      area = math.fabs(area)
      for a in range(len(area_to_sd_map)):
        if area == area_to_sd_map[a]:
          return sign * a / 100
        if 0 < a and area_to_sd_map[a - 1] < area and area < area_to_sd_map[a]:
          # our area is between this value and the previous
          # for simplicity, we will just take the sd half way between a - 1 and a
          return sign * (a - .5) / 100
      return sign * (len(area_to_sd_map) - 1) / 100


# x, y are arrays of sample values
# returns two new arrays with randomly picked
# (with replacement) pairs from x and y
def bootstrap(x, y):
        samp_x = []
    samp_y = []
    num_samples = len(x)
        for i in range(num_samples):
      index = random.randint(0, num_samples - 1)
          samp_x.append(x[index])
        samp_y.append(y[index])
    return (samp_x, samp_y)


def sumofsq(vals, mean):
    count = len(vals)
    total = 0
    for i in range (count):
      diff_sq = (vals[i] - mean)**2
      total += diff_sq
    return total


def sumofproducts(x_vals, y_vals,  mean_x, mean_y):
    count = len(x_vals)
    total = 0
    for i in range (count):
      product = (x_vals[i] - mean_x) * (y_vals[i] - mean_y)
      total += product
    return total


def regressionline(grp_x, grp_y):
    sum_x = sum(grp_x)
    sum_y = sum(grp_y)

    count_x = float(len(grp_x))
    count_y = float(len(grp_y))

    mean_x = sum_x / count_x
    mean_y = sum_y / count_y

    # get the sum of products
    sum_of_prod = sumofproducts(grp_x, grp_y, mean_x, mean_y)
```

```
      # get the sum of squares for x
      sum_of_sq_x = sumofsq(grp_x, mean_x)

      b = sum_of_prod / sum_of_sq_x
      a = mean_y - (b * mean_x)
      return (a, b)


###################################
#
# Computations
#
###################################

# list of lists
samples = []

# file must be in FASTA format
infile=open(input_file)
for line in infile:
        if line.startswith('>'):
                # start of new sample
                samples.append([])
        elif not line.isspace():
                # line must contain values for previous sample
                samples[len(samples) - 1] += map(float,line.split())
infile.close()

grp_x = samples[0]
grp_y = samples[1]

(observed_a, observed_b) = regressionline(grp_x, grp_y)


num_resamples = 10000   # number of times we will resample from our original samples
num_below_observed = 0   # count the number of bootstrap values below the observed
sample statistic
out = []                # will store results of each time we resample

for i in range(num_resamples):
  # bootstrap - then compute our statistic of interest
  # keep pairs together
  # append statistic to out
  (boot_x, boot_y) = bootstrap(grp_x, grp_y)
  # now we have a list of bootstrap samples, run regressionline
  (boot_a, boot_b) = regressionline(boot_x, boot_y)
  if boot_b < observed_b:
    num_below_observed += 1
  out.append(boot_b)

out.sort()

# standard confidence interval computations
tails = (1 - conf_interval) / 2
```

```
# in case our lower and upper bounds are not integers,
# we decrease the range (the values we include in our interval),
# so that we can keep the same level of confidence
lower_bound = int(math.ceil(num_resamples * tails))
upper_bound = int(math.floor(num_resamples * (1 - tails)))


# bias-corrected confidence interval computations
p = num_below_observed / float(num_resamples)# proportion of bootstrap values below
the observed value

dist_from_center = p - .5# if this is negative, the original is below the center, if
positive, it is above
z_0 = area_to_sd(dist_from_center)

# now we want to find the proportion that should be between the mean and one of the
tails
tail_sds = area_to_sd(conf_interval / 2)
z_alpha_over_2 = 0 - tail_sds
z_1_minus_alpha_over_2 = tail_sds

# in case our lower and upper bounds are not integers,
# we decrease the range (the values we include in our interval),
# so that we can keep the same level of confidence
bias_corr_lower_bound = int(math.ceil(num_resamples * (0.5 +
sd_to_area(z_alpha_over_2 + (2 * z_0)))))
bias_corr_upper_bound =  int(math.floor(num_resamples * (0.5 +
sd_to_area(z_1_minus_alpha_over_2 + (2 * z_0)))))


###################################
#
# Output
#
###################################

sign = "+"

if (observed_a < 0):
        sign = "-"

observed_a = abs(observed_a)

print "Line of best fit for observed data: ",
print "y' = %.4f" % observed_b, "x", sign, " %.4f" % observed_a
print "We have", conf_interval * 100, "% confidence that the true slope",
print "is between: %.4f" % out[lower_bound], "and %.4f" % out[upper_bound]

print "***** Bias Corrected Confidence Interval *****"
print "We have", conf_interval * 100, "% confidence that the true slope",
print "is between: %.4f" % out[bias_corr_lower_bound], "and %.4f" %
out[bias_corr_upper_bound]
```

CorrelationSig.py

```
#!/usr/bin/python


######################################
# Linear Correlation Significance Test
# From: Statistics is Easy! By Dennis Shasha and Manda Wilson
#
#
# Assuming that x is not a good predictor of y, tests to see the probability
# of getting a r by chance alone greater than or equal to the one observed (less
# than or equal to the one observed if the observed r is negative).
# Uses shuffling to get a distribution to compare
# to the observed r.
#
# Author: Manda Wilson
#
# Example of FASTA formatted input file:
# >grp_x
# 1350 1510 1420 1210 1250 1300 1580 1310 1290 1320 1490 1200 1360
# >grp_y
# 3.6 3.8 3.7 3.3 3.9 3.4 3.8 3.7 3.5 3.4 3.8 3.0 3.1
#
# Pseudocode:
#
# 1. Calculate r for the observed values (in our example it is .58).
#     a. mean_x = sum of x values / number of x values
#     b. mean_y = sum of y values / number of y values
#     c. Calculate the sum of products:
#        i. Initialize total to 0
#        ii. For each pair of x, y values:
#            I. product = (x - mean_x) * (y - mean_y)
#            II. total += product
#     d. Calculate the sum of squares for the x values:
#        i. Initialize total to 0
#        ii. For each x value:
#            I. diff_sq = (x - mean_x)^2
#            II. total += diff_sq
#     e. Calculate the sum of squares for the y values as we did for the x values in
# step (1d)
#     f. r = sum of products / square root(sum of squares x  * sum of squares y)
#
# 2. Set a counter to 0, this will count the number of times we get a r
#     greater than or equal to 0.58.  Note: if you have a negative r, count
#     the number of times you get a slope less than or equal to the original negative r.
#
# 3. Do the following 10,000 times:
#     a. Shuffle the y values.
#     b. Calculate r on the results from step (3a), just as we did in step (1)
#     c. If r from step (3b) is greater than or equal to our observed r (0.58),
#        increment our counter from step (2).
#
```

```
# 3. counter / 10,000 equals the probability of getting a r greater than or equal to
#      0.58, assuming there is no difference between the groups
#
####################################

import math
import random

####################################
#
# Adjustable variables
#
####################################

input_file = 'input/Correlation.vals'

####################################
#
# Subroutines
#
####################################

def sumofsq(vals, mean):
  count = len(vals)
  total = 0
  for i in range (count):
    diff_sq = (vals[i] - mean)**2
    total += diff_sq
  return total

def sumofproducts(x_vals, y_vals,  mean_x, mean_y):
  count = len(x_vals)
  total = 0
  for i in range (count):
    product = (x_vals[i] - mean_x) * (y_vals[i] - mean_y)
    total += product
  return total

def corrcoef(x, y):
  sum_x = sum(x)
  sum_y = sum(y)
  count_x = float(len(x))
  count_y = float(len(y))
  mean_x = sum_x / count_x
  mean_y = sum_y / count_y

  # get the sum of products
  sum_of_prod = sumofproducts(grp_x, grp_y, mean_x, mean_y)

  # get the sum of squares for x and y
  sum_of_sq_x = sumofsq(grp_x, mean_x)
  sum_of_sq_y = sumofsq(grp_y, mean_y)
```

```
      return sum_of_prod / math.sqrt(sum_of_sq_x * sum_of_sq_y)


###################################
#
# Computations
#
###################################
# list of lists
samples = []

# file must be in FASTA format
infile=open(input_file)
for line in infile:
        if line.startswith('>'):
                # start of new sample
                samples.append([])
        elif not line.isspace():
                # line must contain values for previous sample
                samples[len(samples) - 1] += map(float,line.split())
infile.close()

grp_x = samples[0]
grp_y = samples[1]

observed_r = corrcoef(grp_x, grp_y)


count = 0
num_shuffles = 10000

for i in range(num_shuffles):
   # we want to break the relationship between the pairs, so just shuffle one group
   random.shuffle(grp_y)
   r = corrcoef(grp_x, grp_y)
   if (observed_r > 0 and r >= observed_r) or (observed_r < 0 and r <= observed_r):
     count = count + 1


###################################
#
# Output
#
###################################

print "Observed r: %.2f" % observed_r
print count, "out of 10000 experiments had a r",
if observed_r > 0:
   print "greater than or equal to",
else:
   print "less than or equal to",
print "%.2f" % observed_r
print "Probability that chance alone gave us a r",
```

```
if observed_r > 0:
    print "greater than or equal to",
else:
    print "less than or equal to",
print "%.2f" %observed_r, "is %.2f" % (count / float(num_shuffles))
```

CorrelationConf.py

```
#!/usr/bin/python

######################################
# Linear Correlation Confidence Interval
# From: Statistics is Easy! By Dennis Shasha and Manda Wilson
#
#
# Uses shuffling & bootstrapping to get a 90% confidence interval for the r statistic.
#
# Author: Manda Wilson
#
# Example of FASTA formatted input file:
# >grp_x
# 1350 1510 1420 1210 1250 1300 1580 1310 1290 1320 1490 1200 1360
# >grp_y
# 3.6 3.8 3.7 3.3 3.9 3.4 3.8 3.7 3.5 3.4 3.8 3.0 3.1
#
# Included in the code, but NOT in the pseudocode,
# is the bias-corrected confidence interval.
# See http://www.statisticsiseasy.com/BiasCorrectedConfInter.html
# for more information on bias-corrected cofidence intervals.
#
# Pseudocode:
#
# 1. Calculate r for the observed values (in our example it is .58).
#    a. mean_x = sum of x values / number of x values
#    b. mean_y = sum of y values / number of y values
#    c. Calculate the sum of products:
#       i. Initialize total to 0
#       ii. For each pair of x, y values:
#           I. product = (x - mean_x) * (y - mean_y)
#           II. total += product
#    d. Calculate the sum of squares for the x values:
#       i. Initialize total to 0
#       ii. For each x value:
#           I. diff_sq = (x - mean_x)^2
#           II. total += diff_sq
#    e. Calculate the sum of squares for the y values as we did for the x values in
step (1d)
#    f. r = sum of products / square root(sum of squares x  * sum of squares y)
#
# 2. Do the following 10,000 times:
#    a.
```

```
#        i. Create two new arrays of the same size as the original sample arrays (one
for x and one for y)
#        ii. Fill the new arrays with randomly picked pairs from the original arrays
(randomly picked with replacement)
#    b. Compute the regression line on the bootstrap samples, just as we did in step
(1)
#        with the original samples.
#
# 3. Sort the differences computed in step (2).
#
# 4. Compute the size of each interval tail.  If we want a 90% confidence interval,
then 1 - 0.9 yields the
#    portion of the interval in the tails.  We divide this by 2 to get the size of each
tail, in this case 0.05.
#
# 5. Compute the upper and lower bounds.  To get the lower bound we multiply the tail
size by the number of
#    bootstraps we ran and round this value up to the nearest integer (to make sure
this value maps
#    to an actual boostrap we ran).  In this case we multiple 0.05 by 10,000 and get
500, which rounds up to 500.
#
#    To compute the upper bound we subtract the tail size from 1 and then multiply that
by the number of bootstraps
#    we ran.  Round the result of the last step down to the nearest integer.  Again,
this is to ensure this value
#    maps to an actual bootstrap we ran.  We round the lower bound up and the upper
bound down to reduce the
#    confidence interval size, so that we can still say we have as much confidence in
this result.
#
# 6. The bootstrap values at the lower bound and upper bound give us our confidence
interval.
#
####################################

import math
import random


####################################
#
# Adjustable variables
#
####################################

input_file = 'input/Correlation.vals'
conf_interval = 0.9


####################################
#
# Subroutines
#
####################################
```

```
# maps proportion of values above statistic
# to number of standard deviations above statistic
# keys will be index / 100 \:[0-9]\.[0-9][0-9]\,
area_to_sd_map = [0.0000, 0.0040, 0.0080, 0.0120, 0.0160, 0.0199, 0.0239, 0.0279,
 0.0319, 0.0359, 0.0398, 0.0438, 0.0478, 0.0517, 0.0557, 0.0596, 0.0636, 0.0675,
 0.0714, 0.0753, 0.0793, 0.0832, 0.0871, 0.0910, 0.0948, 0.0987, 0.1026, 0.1064,
 0.1103, 0.1141, 0.1179, 0.1217, 0.1255, 0.1293, 0.1331, 0.1368, 0.1406, 0.1443,
 0.1480, 0.1517, 0.1554, 0.1591, 0.1628, 0.1664, 0.1700, 0.1736, 0.1772, 0.1808,
 0.1844, 0.1879, 0.1915, 0.1950, 0.1985, 0.2019, 0.2054, 0.2088, 0.2123, 0.2157,
 0.2190, 0.2224, 0.2257, 0.2291, 0.2324, 0.2357, 0.2389, 0.2422, 0.2454, 0.2486,
 0.2517, 0.2549, 0.2580, 0.2611, 0.2642, 0.2673, 0.2704, 0.2734, 0.2764, 0.2794,
 0.2823, 0.2852, 0.2881, 0.2910, 0.2939, 0.2967, 0.2995, 0.3023, 0.3051, 0.3078,
 0.3106, 0.3133, 0.3159, 0.3186, 0.3212, 0.3238, 0.3264, 0.3289, 0.3315, 0.3340,
 0.3365, 0.3389, 0.3413, 0.3438, 0.3461, 0.3485, 0.3508, 0.3531, 0.3554, 0.3577,
 0.3599, 0.3621, 0.3643, 0.3665, 0.3686, 0.3708, 0.3729, 0.3749, 0.3770, 0.3790,
 0.3810, 0.3830, 0.3849, 0.3869, 0.3888, 0.3907, 0.3925, 0.3944, 0.3962, 0.3980,
 0.3997, 0.4015, 0.4032, 0.4049, 0.4066, 0.4082, 0.4099, 0.4115, 0.4131, 0.4147,
 0.4162, 0.4177, 0.4192, 0.4207, 0.4222, 0.4236, 0.4251, 0.4265, 0.4279, 0.4292,
 0.4306, 0.4319, 0.4332, 0.4345, 0.4357, 0.4370, 0.4382, 0.4394, 0.4406, 0.4418,
 0.4429, 0.4441, 0.4452, 0.4463, 0.4474, 0.4484, 0.4495, 0.4505, 0.4515, 0.4525,
 0.4535, 0.4545, 0.4554, 0.4564, 0.4573, 0.4582, 0.4591, 0.4599, 0.4608, 0.4616,
 0.4625, 0.4633, 0.4641, 0.4649, 0.4656, 0.4664, 0.4671, 0.4678, 0.4686, 0.4693,
 0.4699, 0.4706, 0.4713, 0.4719, 0.4726, 0.4732, 0.4738, 0.4744, 0.4750, 0.4756,
 0.4761, 0.4767, 0.4772, 0.4778, 0.4783, 0.4788, 0.4793, 0.4798, 0.4803, 0.4808,
 0.4812, 0.4817, 0.4821, 0.4826, 0.4830, 0.4834, 0.4838, 0.4842, 0.4846, 0.4850,
 0.4854, 0.4857, 0.4861, 0.4864, 0.4868, 0.4871, 0.4875, 0.4878, 0.4881, 0.4884,
 0.4887, 0.4890, 0.4893, 0.4896, 0.4898, 0.4901, 0.4904, 0.4906, 0.4909, 0.4911,
 0.4913, 0.4916, 0.4918, 0.4920, 0.4922, 0.4925, 0.4927, 0.4929, 0.4931, 0.4932,
 0.4934, 0.4936, 0.4938, 0.4940, 0.4941, 0.4943, 0.4945, 0.4946, 0.4948, 0.4949,
 0.4951, 0.4952, 0.4953, 0.4955, 0.4956, 0.4957, 0.4959, 0.4960, 0.4961, 0.4962,
 0.4963, 0.4964, 0.4965, 0.4966, 0.4967, 0.4968, 0.4969, 0.4970, 0.4971, 0.4972,
 0.4973, 0.4974, 0.4974, 0.4975, 0.4976, 0.4977, 0.4977, 0.4978, 0.4979, 0.4979,
 0.4980, 0.4981, 0.4981, 0.4982, 0.4982, 0.4983, 0.4984, 0.4984, 0.4985, 0.4985,
 0.4986, 0.4986, 0.4987, 0.4987, 0.4987, 0.4988, 0.4988, 0.4989, 0.4989, 0.4989,
 0.4990, 0.4990]

def sd_to_area(sd):
  sign = 1
  if sd < 0:
    sign = -1
  sd = math.fabs(sd)  # get the absolute value of sd
  index = int(sd * 100)
  if len(area_to_sd_map) <= index:
    return sign * area_to_sd_map[-1] # return last element in array
  if index == (sd * 100):
    return sign * area_to_sd_map[index]
  return sign * (area_to_sd_map[index] + area_to_sd_map[index + 1]) / 2

def area_to_sd(area):
  sign = 1
  if area < 0:
    sign = -1
  area = math.fabs(area)
```

```
   for a in range(len(area_to_sd_map)):
      if area == area_to_sd_map[a]:
         return sign * a / 100
      if 0 < a and area_to_sd_map[a - 1] < area and area < area_to_sd_map[a]:
         # our area is between this value and the previous
         # for simplicity, we will just take the sd half way between a - 1 and a
         return sign * (a - .5) / 100
   return sign * (len(area_to_sd_map) - 1) / 100


# x, y are arrays of sample values
# returns two new arrays with randomly picked
# (with replacement) pairs from x and y
def bootstrap(x, y):
         samp_x = []
   samp_y = []
   num_samples = len(x)
         for i in range(num_samples):
      index = random.randint(0, num_samples - 1)
            samp_x.append(x[index])
         samp_y.append(y[index])
   return (samp_x, samp_y)


def sumofsq(vals, mean):
   count = len(vals)
   total = 0
   for i in range (count):
      diff_sq = (vals[i] - mean)**2
      total += diff_sq
   return total


def sumofproducts(x_vals, y_vals,  mean_x, mean_y):
   count = len(x_vals)
   total = 0
   for i in range (count):
      product = (x_vals[i] - mean_x) * (y_vals[i] - mean_y)
      total += product
   return total


def corrcoef(x, y):
   sum_x = sum(x)
   sum_y = sum(y)
   count_x = float(len(x))
   count_y = float(len(y))
   mean_x = sum_x / count_x
   mean_y = sum_y / count_y

   # get the sum of products
   sum_of_prod = sumofproducts(grp_x, grp_y, mean_x, mean_y)

   # get the sum of squares for x and y
   sum_of_sq_x = sumofsq(grp_x, mean_x)
   sum_of_sq_y = sumofsq(grp_y, mean_y)
```

```
        return sum_of_prod / math.sqrt(sum_of_sq_x * sum_of_sq_y)

#####################################
#
# Computations
#
#####################################

# list of lists
samples = []

# file must be in FASTA format
infile=open(input_file)
for line in infile:
        if line.startswith('>'):
                # start of new sample
                samples.append([])
     elif not line.isspace():
                # line must contain values for previous sample
                samples[len(samples) - 1] += map(float,line.split())
infile.close()

grp_x = samples[0]
grp_y = samples[1]

observed_r = corrcoef(grp_x, grp_y)

num_resamples = 10000   # number of times we will resample from our original samples
num_below_observed = 0   # count the number of bootstrap values below the observed
sample statistic
out = []                 # will store results of each time we resample

for i in range(num_resamples):
  # bootstrap - then compute our statistic of interest
  # keep pairs together
  # append statistic to out
  (boot_x, boot_y) = bootstrap(grp_x, grp_y)
  # now we have a list of bootstrap samples, run corrcoef
  boot_r = corrcoef(boot_x, boot_y)
  if boot_r < observed_r:
    num_below_observed += 1
  out.append(boot_r)

out.sort()

# standard confidence interval computations
tails = (1 - conf_interval) / 2

# in case our lower and upper bounds are not integers,
# we decrease the range (the values we include in our interval),
# so that we can keep the same level of confidence
```

```
lower_bound = int(math.ceil(num_resamples * tails))
upper_bound = int(math.floor(num_resamples * (1 - tails)))

# bias-corrected confidence interval computations
p = num_below_observed / float(num_resamples)# proportion of bootstrap values below
the observed value

dist_from_center = p - .5# if this is negative, the original is below the center, if
positive, it is above
z_0 = area_to_sd(dist_from_center)

# now we want to find the proportion that should be between the mean and one of the
tails
tail_sds = area_to_sd(conf_interval / 2)
z_alpha_over_2 = 0 - tail_sds
z_1_minus_alpha_over_2 = tail_sds

# in case our lower and upper bounds are not integers,
# we decrease the range (the values we include in our interval),
# so that we can keep the same level of confidence
bias_corr_lower_bound = int(math.ceil(num_resamples * (0.5 +
sd_to_area(z_alpha_over_2 + (2 * z_0)))))
bias_corr_upper_bound =  int(math.floor(num_resamples * (0.5 +
sd_to_area(z_1_minus_alpha_over_2 + (2 * z_0)))))

###################################
#
# Output
#
###################################

# print observed value and then confidence interval
print "Observed r: %.2f" % observed_r
print "We have ", conf_interval * 100, "% confidence that the true r",
print "is between: %.2f" % out[lower_bound], "and %.2f" % out[upper_bound]

print "***** Bias Corrected Confidence Interval *****"
print "We have", conf_interval * 100, "% confidence that the true r",
print "is between: %.2f" % out[bias_corr_lower_bound], "and %.2f" %
out[bias_corr_upper_bound]
```

machineANOVAcs.vals

```
>push button
802.5 1018.0 913.0 822.0 868.5 1087.0 530.0 191.0 1070.0 576.0 769.5 598.0 1033.0
861.5 1.0 363.0 650.0 947.0 932.0 907.0 802.5 966.0 844.0 842.5 988.0 1076.0 765.0
652.0 464.5 1061.0 594.0 379.5 590.0 254.0 321.0 46.0 745.0 726.5 373.0 537.5 568.0
586.0 458.0 274.0 143.0 372.0 173.0 430.5 545.0 531.0 471.0 235.0 433.0 298.5 379.5
438.0 713.0 589.0 104.0 222.0 206.0 185.0 155.0 135.0 174.0 343.0 396.5 468.0 110.0
435.0 341.5 269.0 777.0 928.0 889.5 1093.0 917.0 802.5 241.0 1259.0 282.0 242.0 912.0
26.0 324.0 551.0 443.0 13.0 99.0 323.0 712.0 53.0 194.0 773.0 484.0 123.0 8.0 489.0
76.0 486.0 846.0 746.5 840.5 1037.0 998.0 1049.0 970.0 887.0 479.0 977.0 1133.0 379.5
520.0 464.5 839.0 785.0 159.0 508.0 44.0 893.5 631.0 898.0 148.5 461.0 341.5 518.0
722.0 158.0 436.0 305.0 817.5 344.0 203.0 456.0 899.5 463.0 583.5 349.0 971.5 501.0
```

389.0 415.5 290.0 180.0 216.0 168.0 466.5 245.0 817.5 923.5 35.0 262.0 1017.0 357.0
87.0 327.0 1012.0 989.0 1020.0 771.0 742.0 388.0 1148.0 990.0 653.5 868.5 500.0 742.0
1259.0 1043.5 829.0 971.5 671.0 918.0 552.0 1060.0 258.0 136.0 325.0 382.5 946.0 140.0
619.5 714.5 742.0 716.5 394.0 661.0 544.0 996.0 294.5 499.0 541.0 856.5 748.0 795.0
726.5 311.0 447.5 813.5 407.0 674.0 574.0 566.0 439.5 623.0 374.5 629.5 543.0 404.0
177.0 427.5 516.0 555.5 997.0 85.0 469.0 255.0 444.0 945.0 974.0 832.0 847.5 887.0
786.5 414.0 933.5 362.0 762.5 1084.0 1055.0 664.0 767.0 714.5 1047.5 639.0 987.0 737.0
595.0 353.0 935.5 677.0 558.0 835.0 423.0 176.0 651.0 754.0 950.5 772.0 740.0 958.0
746.5 686.0 588.0 691.0 400.0 206.0 757.0 396.5 453.5 1043.5 751.0 1050.5 667.5 528.0
466.5 447.5 849.0 613.0 948.0 726.5 625.0 809.0 1007.0 939.0 831.0 867.0 1073.0 980.0
346.5 797.5 260.0 732.0 401.0 525.0 762.5 314.5 413.0 782.0 496.5 502.5 326.0 350.0
493.0 367.0 909.0 1011.0 978.0 923.5 807.5 619.5 445.0 1000.0 875.0 1000.0 188.0 470.0
665.0 979.0 694.5 491.0 711.0 941.5 564.5 310.0 168.0 1128.5 943.0 1013.0 733.0 802.5
1052.0 639.0 792.0 825.0 504.0 822.0 289.0 623.0 833.0 777.0 906.0 860.0 827.0 834.0
689.0 1063.0 864.0 409.5 781.0 786.5 409.5 908.0 810.0 891.5 546.0 629.5 692.0 855.0
784.0 402.5 506.5 575.0 790.0 1071.5 995.0 644.0 853.5 560.5 494.5 549.0 902.0 1000.0
859.0 679.0 700.0 984.5 889.5 377.0 612.0 580.5 462.0 720.0 838.0 872.5 616.0 577.0
309.0 596.5 567.0 430.5 599.5 676.0 830.0 850.0 840.5 98.0 637.0 847.5 526.0 961.0
885.0 935.5 656.5 872.5 1134.0 954.0 460.0 736.0 293.0 506.5 1081.0 557.0 975.0 1259.0
267.5 758.0 895.0 603.0 791.0 813.5 793.0 83.0 89.0 109.0 458.0 128.0 105.5 23.0 134.0
183.0 306.0 118.0 836.0 116.5 1095.5 15.0 63.0 71.0 949.0 33.0 548.0 334.0 61.0 369.0
336.0 387.0 472.0 48.0 374.5 1057.5 477.0 165.0 498.0 58.5 80.0 4.0 55.0 137.0 596.5
1043.5 365.5 151.5 296.0 626.0 723.0 851.0 257.0 175.0 56.0 228.5 659.0 22.0 828.0
632.0 215.0 300.0 351.0 371.0 1005.0 914.0 447.5 278.0 308.0 635.0 901.0 494.5 60.0
54.0 802.5 1067.0 666.0 563.0 706.0 510.5 346.5 1071.5 94.0 716.5 780.0 125.5 405.0
88.0 333.0 285.5 455.0 331.0 510.5 688.0 923.5 878.0 957.0 933.5 1002.0 963.0 903.0
360.5 569.0 937.0 3.0 1023.0 816.0 926.0 609.5 142.0 521.0 882.0 984.5 1019.0 967.0
941.5 842.5 726.5 198.0 617.0 754.0 1026.0 562.0 684.0 301.0 591.5 285.5 883.5 1015.5
643.0 802.5 702.0 871.0 515.0 1003.0 669.0 369.0 209.0 513.0 1099.0 682.5 450.0 927.0
667.5 964.5 709.0 994.0 865.5 1046.0 615.0 982.0 537.5 231.0 1259.0 532.5 318.0 1259.0
439.5 392.0 121.0 232.0 492.0 411.5 744.0 570.5 230.0 502.5 703.5 642.0 1259.0 93.0
112.0 332.0 233.5 6.5 58.5 279.0 131.0 181.0 66.0 247.0 38.0 266.0 458.0 245.0 78.5
338.0 243.0 303.0 693.0 442.0 1259.0 1259.0 1162.5 789.0 1259.0 1009.5 1259.0 1153.5
1059.0 1259.0 1102.0 1259.0 959.0 1078.0 1141.0 1160.0 1123.5 1143.0 1171.5 1054.0
1119.5 1156.5 1066.0 1167.0 1259.0 1095.5 981.0 1098.0 1190.0 1125.5 1110.0 1035.0
1092.0 1075.0 1056.0 954.0 1259.0 1028.0 954.0 1085.0 1074.0 1039.0 1169.0 395.0 132.5
281.0 154.0 25.0 30.0 560.5 162.0 208.0 187.0 153.0 160.0 168.0 119.5 984.5 256.0
386.0 307.0 313.0 39.0 171.0 195.0 170.0 559.0 658.0 12.0 16.0 6.5 5.0 294.5 64.0
189.0 192.0 298.5 32.0 105.5 609.5 193.0 240.0 356.0 678.0 437.0 819.0 752.0 761.0
190.0 125.5 815.0 101.0 302.0 402.5 291.0 68.0 145.0 97.0 239.0 119.5 67.0 214.0 976.0
20.0 182.0 164.0 31.0 822.0 280.0 211.0 984.5 481.0 837.0 639.0 822.0 876.0 767.0
921.0 287.0 583.5 522.0 681.0 236.0 359.0 379.5 675.0 259.0 84.0 364.0 390.0 478.0
224.5 730.0 896.5 264.0 475.5 883.5 627.0 138.0 422.0 578.0 962.0 421.0 249.0 220.5
750.0 671.0 210.0 369.0 1113.0 418.0 519.0 554.0 312.0 427.5 673.0 317.0 861.5 340.0
735.0 635.0 425.0 115.0 923.5 212.0 845.0 72.0 705.0 487.0 237.0 148.5 1024.5 250.0
475.5 417.0 196.0 1259.0 682.5 645.0 482.5 69.0 1259.0 707.0 759.5 184.0 452.0 788.0
200.0 273.0 719.0 161.0 166.0 201.0 147.0 45.0 78.5 73.5 335.0 57.0 82.0 144.0 179.0
2.0 655.0 21.0 858.0 14.0 50.0 157.0 51.0 17.0 65.0 178.0 186.0 29.0 77.0 130.0 219.0
124.0 252.0 599.5 641.0 277.0 129.0 95.0 103.0 297.0 863.0 756.0 473.0 96.0 75.0 517.0
529.0 879.0 424.0 34.0 204.0 163.0 43.0 28.0 92.0 36.0 292.0 490.0 146.0 731.0 607.0
701.0 352.0 919.0 779.0 365.5 775.0 685.0 1024.5 663.0 696.0 697.0 512.0 954.0 419.0
826.0 767.0 587.0 555.5 870.0 532.5 1077.0 853.5 991.0 1259.0 806.0 1014.0 799.0 330.0
865.5 662.0 856.5 633.0 283.0 252.0 572.5 411.5 729.0 19.0 1021.0 579.0 284.0 896.5
322.0 542.0 796.0 393.0 1047.5 1100.0 271.0 1033.0 1027.0 653.5 724.0 954.0 623.0
794.0 86.0 797.5 582.0 537.5 968.0 1004.0 660.0 102.0 916.0 606.0 496.5 591.5 920.0
807.5 602.0 881.0 384.5 355.0 734.0 238.0 1127.0 964.5 227.0 915.0 267.5 488.0 671.0
687.0 739.0 759.5 929.0 127.0 339.0 272.0 524.0 328.5 474.0 398.5 992.0 328.5 580.5

441.0 376.0 969.0 939.0 764.0 904.0 480.0 893.5 570.5 354.0 223.0 939.0 777.0 564.5
618.0 1259.0 505.0 1033.0 874.0 880.0 905.0 527.0 635.0 547.0 822.0 248.0 703.5 453.5
304.0 451.0 213.0 960.0 930.0 151.5 648.0 601.0 52.0 656.5 111.0 226.0 49.0 585.0
1259.0 415.5 108.0 391.0 113.0 346.5 228.5 316.0 220.5 11.0 24.0 114.0 42.0 27.0 360.5
206.0 139.0 224.5 141.0 270.0 18.0 62.0 156.0 430.5 265.0 202.0 90.0 41.0 172.0 70.0
609.5 122.0 37.0 10.0 40.0 81.0 100.0 252.0 116.5 47.0 1259.0 9.0 406.0
>touch screen
1009.5 107.0 680.0 710.0 261.0 482.5 649.0 605.0 593.0 550.0 698.0 754.0 811.5 245.0
430.5 319.0 263.0 1029.5 718.0 540.0 346.5 899.5 217.5 783.0 604.0 420.0 609.5 509.0
694.5 621.0 699.0 877.0 1259.0 769.5 628.0 1130.5 1179.0 1031.0 1147.0 1150.0 1196.0
1186.0 1188.0 1259.0 1132.0 1065.0 1192.0 1050.5 1101.0 1108.5 1164.5 1146.0 1116.0
1259.0 1007.0 1183.0 1053.0 1259.0 1086.0 1181.0 1194.0 1197.0 1068.5 1176.0 1080.0
1144.0 1259.0 1149.0 1184.0 1168.0 1259.0
>optical scan
1161.0 1068.5 1105.0 1259.0 1259.0 1114.0 1259.0 1259.0 514.0 1259.0 910.5 993.0 852.0
973.0 485.0 891.5 1038.0 721.0 749.0 1007.0 434.0 314.5 1123.5 217.5 931.0 708.0 275.0
572.5 426.0 944.0 535.0 523.0 233.5 288.0 614.0 337.0 887.0 534.0 91.0 358.0 384.5
646.5 320.0 738.0 132.5 646.5 537.5 950.5 447.5 553.0 910.5 197.0 199.0 382.5 811.5
690.0 398.5 774.0 1041.0 1259.0 1259.0 1082.0 1137.0 1259.0 1119.5 1259.0 1121.5
1259.0 1112.0 1259.0 1259.0 1259.0 1259.0 1259.0 1259.0 1259.0 1259.0 1182.0 1115.0
1108.5 1259.0 1259.0 1259.0 1259.0 1259.0 1259.0 1090.0 1259.0 1259.0 1259.0 1259.0
1175.0 1062.0 1259.0 1259.0 1259.0 1036.0 150.0 276.0 408.0 1259.0 1191.0 1125.5
1259.0 1259.0 1180.0 1151.0 1164.5 1174.0 1089.0 1259.0 1259.0 1178.0 1259.0 1259.0
1135.5 1088.0 1259.0 1259.0 1259.0 1015.5 1259.0 1259.0 1111.0 1259.0 1177.0 1259.0
1259.0 1155.0 1097.0 1259.0 1170.0 1259.0 1259.0 1029.5 1259.0 1259.0 1166.0 1259.0
1145.0 1079.0 1259.0 1153.5 1259.0 1158.0 1187.0 1259.0 1043.5 1259.0 1259.0 1259.0
1259.0 1173.0 1118.0 1195.0 1162.5 1259.0 1259.0 1259.0 1121.5 1259.0 1083.0 1259.0
1259.0 1103.0 1259.0 1107.0 1259.0 1259.0 1259.0 1117.0 1259.0 1128.5 1057.5 1259.0
1022.0 1106.0 1193.0 1259.0 1259.0 1259.0 1142.0 1259.0 1159.0 1156.5 1259.0 1259.0
1259.0 1259.0 1259.0 1139.5 1259.0 1259.0 1171.5 1259.0 1189.0 1259.0 1130.5 1138.0
1259.0 1091.0 1185.0 1259.0 1259.0 1259.0 1259.0 1259.0 1135.5 1259.0 1259.0 1139.5
1259.0 1152.0 1094.0 1259.0 1259.0 1104.0 73.5 1259.0 1040.0 1064.0 1259.0

push_button.vals

>percent minority
100.0 99.9 99.88 99.85 99.8 99.8 99.79 99.7 99.67 99.66 99.65 99.62 99.48 99.45 99.43
99.4 99.39 99.37 99.25 99.18 99.15 99.1 98.96 98.93 98.85 98.73 98.59 98.51 98.49
98.48 98.47 98.43 98.42 98.4 98.36 98.28 98.13 98.1 98.08 98.08 98.01 89.73 89.71 89.7
89.56 89.45 89.27 89.22 89.19 89.15 89.08 89.07 88.99 88.85 88.84 88.78 88.76 88.74
88.7 88.68 88.66 88.6 88.55 88.41 88.28 88.17 88.16 88.09 88.03 87.58 87.53 87.43
87.42 87.37 87.32 87.23 87.19 87.07 87.01 86.98 86.97 86.9 86.88 86.81 86.78 86.68
86.62 86.53 86.49 86.48 86.44 86.44 86.36 86.34 86.26 86.24 86.21 86.07 85.89 85.89
85.88 85.71 85.61 85.51 85.41 85.39 85.18 85.06 84.94 84.77 84.55 84.49 84.43 84.26
84.23 84.18 84.13 83.99 83.96 83.88 83.85 83.84 83.77 83.76 83.67 83.67 83.64 83.59
83.29 82.8 82.73 82.7 82.56 82.56 82.53 82.5 82.42 82.32 82.29 82.05 82.04 81.99 81.84
81.72 81.61 81.61 81.45 81.4 81.3 81.29 81.26 81.24 81.22 81.02 80.95 80.95 80.95
80.87 80.86 80.76 80.71 80.3 80.21 80.17 80.17 97.91 97.87 97.86 97.76 97.68 97.68
97.62 97.51 97.38 97.33 97.32 97.28 97.07 79.91 79.9 79.76 79.7 79.69 79.58 79.51
79.39 79.07 79.07 79.02 79.01 78.78 78.7 78.64 78.51 78.45 78.35 78.24 78.17 78.13
78.11 78.03 78.0 77.9 77.87 77.81 77.78 77.42 77.29 76.98 76.93 76.9 76.78 76.73 76.7
76.69 76.66 76.54 76.21 76.12 76.11 76.1 76.02 75.86 75.8 75.77 75.63 75.61 75.58
75.43 75.38 75.26 75.18 75.1 75.06 74.89 74.87 74.85 74.83 74.57 74.33 74.29 74.1 74.0
73.71 73.63 73.42 73.38 73.28 73.19 72.77 72.76 72.61 72.55 72.42 72.36 72.32 72.17
71.99 71.89 71.87 71.75 71.73 71.73 71.72 71.63 71.58 71.56 71.5 71.36 71.36 71.35
71.22 71.08 70.96 70.95 70.89 70.88 70.83 70.06 70.01 96.74 96.67 96.63 96.4 96.13

69.96 69.92 69.9 69.88 69.86 69.78 69.43 69.23 69.16 69.15 69.11 69.07 68.99 68.93
68.92 68.88 68.84 68.57 68.57 68.21 68.16 68.14 67.92 67.78 67.62 67.53 67.44 67.34
67.32 67.3 67.26 67.16 67.01 67.01 66.96 66.89 66.86 66.84 66.58 66.56 66.43 66.32
66.16 66.15 66.02 66.01 65.98 65.83 65.81 65.56 65.52 65.51 65.45 65.28 65.28 65.15
64.74 64.72 64.61 64.55 64.52 64.5 64.5 64.44 64.42 64.37 64.33 64.29 64.23 64.04
63.87 63.73 63.65 63.51 63.5 63.44 63.39 63.39 63.29 63.2 63.19 63.17 63.11 62.85
62.71 62.68 62.66 62.62 62.54 62.32 62.3 62.2 62.16 62.03 61.65 61.5 61.24 61.19 60.78
60.72 60.72 60.67 60.51 60.43 60.41 60.34 60.32 60.21 95.73 95.54 95.46 59.96 59.79
59.74 59.54 59.49 59.14 59.1 58.83 58.19 58.17 57.97 57.93 57.89 57.89 57.87 57.73
57.71 57.71 57.66 57.46 57.41 57.0 56.75 56.4 56.12 56.09 55.95 55.92 55.91 55.84
55.83 55.8 55.78 55.7 55.41 55.33 55.11 55.08 55.04 54.94 54.93 54.87 54.84 54.76
54.55 54.39 54.24 54.12 53.71 53.69 53.44 53.4 53.37 53.37 53.33 53.24 53.23 53.18
53.02 53.0 52.91 52.89 52.86 52.67 52.47 52.44 52.42 52.28 52.28 52.27 52.24 52.2
52.18 52.12 52.08 51.97 51.89 51.75 51.53 51.47 51.35 51.27 51.24 51.17 51.09 51.03
50.86 50.86 50.79 50.7 50.64 50.62 50.61 50.43 50.26 50.18 50.17 94.93 94.89 94.63
94.63 94.61 94.59 94.58 94.56 94.5 94.26 94.14 50.0 49.84 49.69 49.68 49.68 49.54
49.52 49.4 49.25 49.2 49.18 49.1 49.07 49.06 48.9 48.9 48.89 48.8 48.7 48.49 48.47
48.46 48.31 48.29 48.21 48.2 48.15 48.13 47.94 47.92 47.87 47.85 47.82 47.82 47.61
47.6 47.57 47.41 47.4 47.26 47.11 47.04 46.68 46.65 46.62 46.59 46.5 46.5 46.28 46.08
46.05 46.03 45.96 45.78 45.71 45.55 45.36 45.3 45.22 45.19 45.19 45.02 44.93 44.93
44.86 44.81 44.78 44.64 44.62 44.51 44.44 44.41 44.39 44.37 44.36 44.34 44.33 44.3
44.29 44.19 44.15 44.13 44.12 44.01 43.93 43.89 43.86 43.8 43.8 43.73 43.69 43.61
43.51 43.49 43.48 43.47 43.46 43.32 43.29 43.19 43.1 43.03 42.99 42.84 42.83 42.77
42.73 42.72 42.71 42.68 42.53 42.47 42.44 42.16 42.11 42.1 42.05 41.96 41.94 41.94
41.93 41.9 41.9 41.86 41.81 41.77 41.73 41.7 41.67 41.59 41.52 41.34 41.2 41.04 41.0
40.96 40.88 40.71 40.66 40.63 40.43 40.22 40.13 40.07 93.86 93.77 93.73 93.61 93.44
93.33 93.16 93.11 39.95 39.86 39.82 39.7 39.67 39.55 39.45 39.4 39.34 39.33 39.31
39.14 39.08 39.03 38.99 38.97 38.97 38.92 38.91 38.84 38.74 38.72 38.69 38.54 38.52
38.49 38.48 38.47 38.3 38.28 38.25 38.23 38.11 38.1 38.03 37.91 37.91 37.8 37.73 37.67
37.55 37.5 37.48 37.41 37.3 37.22 37.17 37.14 37.06 36.97 36.91 36.85 36.76 36.75
36.69 36.63 36.6 36.56 36.5 36.49 36.37 36.36 36.24 36.22 36.2 36.19 36.1 36.05 36.04
36.04 36.01 35.96 35.92 35.92 35.91 35.84 35.83 35.71 35.39 35.38 35.36 35.33 35.1
34.98 34.9 34.82 34.82 34.81 34.79 34.69 34.58 34.55 34.51 34.43 34.38 34.26 34.03
33.97 33.97 33.95 33.94 33.92 33.91 33.89 33.8 33.46 33.33 33.23 33.16 33.14 33.07
33.01 32.72 32.65 32.57 32.48 32.35 32.33 32.27 32.22 32.21 32.2 32.2 32.09 32.04 31.9
31.76 31.74 31.71 31.57 31.51 31.38 31.34 31.28 31.26 31.2 31.2 31.19 31.17 31.16
31.11 31.09 31.08 31.08 31.07 30.92 30.9 30.81 30.68 30.63 30.58 30.54 30.47 30.43
30.41 30.39 30.31 30.2 30.19 30.16 92.95 92.78 92.62 92.51 92.45 92.42 92.41 92.3
92.26 92.11 92.06 92.05 29.92 29.66 29.55 29.49 29.47 29.34 29.26 29.17 29.06 29.02
28.95 28.93 28.93 28.85 28.8 28.8 28.76 28.7 28.67 28.62 28.54 28.48 28.4 28.39 28.38
28.33 28.32 28.24 28.23 28.03 28.0 27.87 27.8 27.75 27.67 27.64 27.55 27.49 27.48
27.34 27.29 27.25 27.02 26.97 26.56 26.56 26.53 26.47 26.44 26.34 26.26 26.11 26.08
26.06 25.91 25.87 25.79 25.64 25.5 25.47 25.42 25.37 25.34 25.31 25.31 25.17 25.16
25.07 25.0 24.96 24.95 24.88 24.81 24.76 24.69 24.58 24.55 24.43 24.4 24.35 24.33
24.28 24.23 24.19 24.17 23.93 23.86 23.74 23.48 23.19 23.07 23.01 22.99 22.97 22.97
22.91 22.89 22.59 22.48 22.43 22.37 22.36 22.3 22.24 22.18 22.14 22.06 22.01 21.95
21.84 21.83 21.77 21.74 21.74 21.73 21.73 21.6 21.57 21.56 21.56 21.44 21.43 21.37
21.37 21.35 21.31 21.15 21.05 21.04 20.86 20.75 20.69 20.5 20.44 20.43 20.29 20.11
20.07 20.03 91.98 91.96 91.88 91.44 91.43 91.35 91.33 91.28 91.27 91.16 91.06 19.84
19.77 19.77 19.5 19.39 19.38 19.17 19.17 19.07 18.99 18.91 18.9 18.86 18.86 18.78
18.44 18.41 18.26 18.18 18.08 18.01 17.97 17.94 17.83 17.81 17.61 17.54 17.51 17.42
17.39 17.35 17.33 17.28 17.25 17.07 17.03 16.92 16.9 16.84 16.75 16.67 16.52 16.35
16.25 15.92 15.84 15.68 15.56 15.45 15.33 15.31 15.04 14.53 14.52 14.09 13.96 13.87
13.86 13.67 12.77 11.99 11.79 11.63 10.93 10.81 10.69 10.61 10.28 10.23 90.93 90.91
90.9 90.8 90.77 90.75 90.74 90.73 90.73 90.68 90.62 90.36 90.24 90.19 90.09 90.09
90.09 9.95 8.52 8.29 7.85 4.44 0.76

```
>percent undervote
8.0 4.69 9.65 9.93 10.52 6.88 5.62 7.78 8.52 9.76 4.44 16.67 13.05 5.58 12.63 6.06
10.34 14.17 6.69 9.78 4.0 12.65 5.97 7.75 13.59 9.26 4.74 13.84 5.13 8.04 5.52 4.15
10.18 12.82 16.05 5.42 6.56 6.81 5.13 9.97 2.23 7.45 9.39 4.09 4.65 4.21 6.95 7.14
7.84 7.97 4.81 7.57 14.71 12.45 14.75 11.7 3.76 7.79 6.45 6.72 5.77 10.91 6.85 8.98
8.27 8.79 9.09 7.12 6.01 11.57 19.79 9.02 10.42 8.99 4.09 5.39 8.52 4.31 10.78 5.36
4.56 9.19 5.97 8.82 6.18 6.64 4.89 11.18 9.76 4.73 3.51 8.68 7.69 4.63 5.5 5.91 8.42
7.75 5.36 9.63 4.02 6.97 5.67 11.15 6.3 5.94 6.73 20.55 9.19 7.21 5.28 9.26 8.46 1.56
4.42 5.99 12.99 5.0 10.42 6.31 7.75 6.71 12.12 4.23 5.24 8.24 8.77 11.31 1.75 10.26
11.63 5.56 7.21 9.09 7.73 9.38 16.67 8.22 13.56 7.55 6.99 5.49 8.09 0.0 13.47 5.44
6.32 8.64 9.62 7.14 8.94 0.98 6.2 5.48 2.05 3.89 8.74 8.83 4.29 4.58 5.71 7.25 14.58
10.1 7.56 7.41 14.91 14.18 6.79 12.62 6.97 5.86 6.09 5.4 9.95 17.3 9.23 8.72 15.42
6.38 11.32 10.71 9.95 8.23 15.05 6.91 4.58 6.92 11.42 6.06 6.78 6.76 6.38 4.72 8.33
1.49 3.85 7.5 5.33 5.47 14.29 5.68 3.94 5.18 8.86 10.9 7.0 1.26 11.22 4.41 5.84 4.5
7.55 4.38 3.8 8.21 2.2 9.86 8.74 4.4 0.0 5.59 3.11 10.34 8.21 6.87 11.18 6.92 9.57
7.14 6.86 5.29 10.08 8.81 7.88 0.0 8.64 6.18 8.06 5.0 3.83 10.55 4.95 2.38 4.64 5.32
6.37 11.48 8.51 5.05 4.91 5.91 9.24 3.19 4.84 5.08 3.91 6.13 4.96 6.41 0.0 6.38 8.19
2.29 6.3 4.69 10.93 14.01 4.49 7.73 9.68 7.11 4.49 1.36 4.44 3.95 10.08 8.77 2.27
11.21 6.85 6.5 13.36 8.72 3.95 9.34 6.22 3.15 0.84 4.09 7.22 12.18 7.23 4.41 1.7 5.88
8.28 6.22 4.98 4.43 4.42 6.59 3.7 7.56 8.33 1.32 4.95 2.53 26.98 5.14 3.51 6.38 6.75
6.33 7.11 2.78 10.67 6.12 6.85 6.49 3.48 7.78 5.37 4.07 1.46 2.52 3.52 14.06 4.85 8.66
6.53 1.79 6.54 0.0 4.27 3.23 2.67 1.04 8.94 9.67 37.13 5.75 6.41 6.42 1.36 6.22 10.12
7.01 7.41 3.93 6.29 4.74 1.4 3.95 5.05 7.06 4.61 2.46 4.55 3.95 2.47 15.25 3.63 12.05
5.16 4.24 8.81 7.1 6.99 4.94 7.94 5.21 2.01 3.47 3.42 7.09 6.34 1.0 4.46 4.19 2.84
3.37 6.4 4.06 6.98 7.09 9.39 5.42 3.13 3.66 11.54 4.74 1.86 5.9 7.16 5.61 4.94 5.04
7.4 9.42 5.36 3.16 0.99 5.97 7.89 4.96 1.99 5.41 0.0 10.17 10.15 3.45 3.38 6.03 3.59
2.99 3.72 0.0 4.44 7.45 3.75 4.58 9.52 4.9 4.84 4.29 4.52 5.04 3.66 3.47 5.68 6.19
3.02 6.13 5.14 2.84 6.01 6.31 2.43 3.91 5.59 5.71 5.83 4.32 6.0 4.91 2.5 6.52 2.91 7.1
1.01 1.0 7.74 3.45 6.25 4.51 2.38 3.45 3.56 2.93 8.92 4.41 4.5 10.92 7.34 3.15 8.88
2.33 6.36 3.3 7.48 5.75 3.8 4.14 11.11 2.17 2.63 5.07 5.49 5.56 4.13 9.06 3.6 3.64
2.97 1.66 6.62 4.08 6.43 5.41 9.96 5.06 4.2 8.32 6.51 10.11 7.11 9.77 7.91 8.26 10.51
7.31 13.68 10.3 2.49 4.23 5.63 1.36 4.65 0.0 7.43 7.43 2.17 2.63 4.44 5.3 1.26 1.3
2.33 5.17 6.14 5.48 5.05 2.05 4.37 4.44 8.84 5.56 5.32 4.26 0.4 2.45 7.69 3.77 2.6
5.36 4.59 6.53 3.85 2.4 4.17 3.66 5.09 5.15 3.24 1.83 1.42 1.71 5.29 3.61 3.46 4.66
6.21 4.9 1.22 4.89 0.46 6.38 5.11 1.8 4.6 2.82 3.91 2.34 2.49 4.05 1.13 5.33 3.52 3.69
0.88 3.32 4.58 2.0 3.74 3.02 3.72 4.51 2.45 2.3 10.8 5.32 4.64 4.17 2.61 6.05 4.77
3.17 10.68 5.62 0.65 2.32 2.58 7.35 4.24 5.96 8.93 2.79 2.8 1.63 2.38 2.77 4.82 3.0
4.14 5.92 5.63 2.31 3.31 2.28 0.0 2.58 0.0 3.67 5.47 3.8 4.38 5.31 4.23 3.57 3.25 1.8
2.52 2.96 7.21 1.71 3.79 3.96 4.61 0.96 2.05 3.82 5.26 4.99 4.24 1.97 3.75 2.35 2.87
4.11 13.7 2.66 1.65 3.45 4.81 2.34 3.16 2.73 8.51 6.4 3.67 3.03 11.22 2.17 8.96 7.66
1.85 5.22 3.0 2.51 1.62 4.22 1.76 2.88 0.0 2.12 6.64 3.6 3.99 2.17 4.67 5.87 9.02 3.38
1.93 4.03 7.31 5.38 1.49 4.26 5.71 3.35 0.0 6.48 3.86 3.45 5.9 5.07 3.14 1.75 3.14
4.42 4.57 4.9 3.58 3.47 5.98 3.73 4.66 4.67 3.55 4.33 5.85 5.73 5.21 4.08 3.57 2.26
3.88 3.56 3.96 2.33 8.25 4.17 9.43 1.56 2.92 1.01 5.66 4.34 7.04 2.65 1.18 3.01 2.88
4.97 0.0 1.52 3.17 3.37 2.8 2.04 2.66 1.47 4.31 2.2 1.3 6.85 3.29 0.72 2.37 12.64 2.63
4.46 3.45 4.72 3.14 5.83 1.28 1.47 3.09 3.79 1.23 4.6 8.9 3.76 4.11 3.93 3.47 5.03
5.04 4.14 4.2 7.26 5.23 3.1 3.53 4.4 5.11 1.66 4.41 2.44 5.9 3.09 3.25 2.38 2.05 1.9
3.0 4.28 2.2 3.1 6.0 2.04 3.57 4.0 2.17 4.21 5.97 1.54 4.58 4.43 5.38 5.78 0.29 3.99
4.41 2.87 3.83 5.41 3.7 2.72 1.65 4.04 4.44 2.22 2.91 7.6 3.56 1.85 2.92 6.75 1.96
4.35 3.43 5.71 7.9 6.25 8.15 10.53 5.31 4.7 7.63 4.24 14.39 11.24 6.01 7.95 0.79 0.0
3.76 3.03 3.42 3.02 3.76 4.62 5.23 1.57 0.0 1.51 1.99 3.14 0.0 2.84 4.4 1.56 4.81 4.15
5.25 6.98 5.12 3.85 2.49 3.4 1.46 5.28 2.08 7.18 7.69 3.7 4.57 3.61 2.63 3.5 0.67 4.41
2.7 2.9 8.38 1.37 4.69 3.33 5.39 6.06 3.1 2.29 4.51 2.11 0.54 7.98 4.5 4.22 6.21 3.5
1.98 3.42 2.67 5.41 4.48 6.8 3.8 1.98 3.18 3.97 4.55 4.35 3.37 4.71 2.33 5.73 2.75
0.58 3.33 3.52 0.38 1.34 4.03 2.75 3.97 7.42 0.56 2.91 0.53 4.07 6.73 5.19 3.85 3.28
1.51 3.33 0.97 1.66 2.85 4.26 3.2 2.12 3.01 2.36 2.66 0.0 2.2 7.48 1.36 2.07 2.5 3.61
0.42 2.51 1.73 4.36 2.97 3.26 2.36 5.08 3.79 2.72 0.63 1.53 3.28 3.05 1.49 13.04 2.98
```

4.62 3.96 5.03 1.81 2.4 3.31 1.29 1.69 5.8 2.91 2.93 6.03 4.78 4.24 8.33 4.37 12.26
4.95 5.02 8.37 2.63 6.72 8.31 8.27 5.67 1.53 1.17 3.92 1.38 3.6 4.72 2.87 3.01 1.93
3.38 3.5 6.9 2.38 5.03 1.5 2.23 1.22 9.6 3.92 4.6 5.47 7.37 0.69 6.99 3.27 2.51 2.28
5.34 0.59 0.0 2.79 3.05 2.45 7.18 3.32 3.91 3.85 4.61 2.5 2.84 1.3 4.37 0.51 2.04 7.58
2.38 1.2 2.8 6.75 2.11 2.58 12.5 3.2 3.9 6.14 2.29 7.2 0.7 4.82 2.95 1.66 4.85 0.0
2.95 0.84 3.33 0.0 3.76 4.85 9.88 10.76 13.32 2.92 9.62 5.96 5.44 6.48 8.8 5.85 6.03
4.67 11.44 6.77 3.08 6.3 8.23 2.98 5.19 3.67 0.73 4.88 2.17

optical.vals

>percent minority
89.23 88.26 87.76 87.14 86.56 83.96 83.45 82.7 82.5 80.82 79.2 79.14 79.12 79.1 77.84
77.53 75.36 73.88 73.24 72.95 72.86 72.61 71.49 71.16 69.51 68.54 68.33 65.4 64.76
63.33 63.29 63.15 62.5 62.27 62.27 61.98 61.94 61.91 61.65 61.13 60.42 59.45 58.75
58.6 58.55 58.31 57.54 57.47 56.49 56.13 55.76 55.37 55.09 54.89 54.73 54.56 54.35
54.32 51.45 50.28 50.0 50.0 49.69 48.46 47.9 47.58 47.06 46.68 46.5 46.25 45.6 45.56
45.43 45.32 45.24 44.97 44.72 43.99 43.18 42.88 42.63 42.63 42.44 42.42 41.63 41.08
40.85 40.53 40.28 40.24 40.09 40.04 39.68 39.49 39.42 39.4 39.06 38.76 38.45 38.14
37.99 37.46 37.44 37.13 36.8 36.75 36.63 36.23 36.11 35.71 34.99 34.47 33.33 33.27
33.09 33.01 32.6 31.99 31.88 31.53 31.37 31.3 30.69 30.42 30.27 29.04 28.95 28.56
27.86 27.76 26.74 26.6 26.52 26.28 25.86 25.77 25.54 25.44 25.33 25.31 25.12 23.96
23.86 23.8 23.77 23.29 23.19 23.18 22.97 22.81 22.39 22.11 21.86 21.8 21.78 21.69
21.62 21.52 21.35 21.13 21.11 20.77 20.71 20.67 20.32 91.43 91.33 19.9 19.86 19.57
19.43 19.14 18.48 18.12 18.1 17.97 17.69 17.58 17.52 17.52 17.45 17.1 16.86 16.84
16.03 15.71 15.52 15.39 15.38 15.32 14.95 14.72 14.46 13.69 13.67 13.63 13.22 12.01
11.93 11.43 11.32 10.39 10.31 10.12 9.87 9.84 9.62 9.47 8.53 8.24 8.02 7.9 7.69 4.82
1.49
>percent undervote
0.0 0.79 0.0 0.0 4.3 0.0 0.27 0.3 6.46 0.67 0.51 0.0 0.0 10.0 1.06 0.0 0.93 5.74 1.04
2.9 4.86 0.0 0.3 6.15 1.05 0.0 0.92 0.78 3.59 4.63 8.96 0.0 0.0 0.0 7.5 4.3 1.79 0.0
0.33 0.0 5.84 4.05 0.0 0.0 4.86 0.37 0.0 7.34 0.0 0.0 0.72 0.83 0.0 1.61 0.0 7.79 2.48
5.96 7.76 0.81 0.0 4.85 5.5 4.94 0.0 0.54 6.95 0.63 0.0 0.91 0.38 0.0 0.32 0.0 0.36
0.0 0.54 6.8 1.67 0.0 0.0 5.6 3.39 1.86 0.0 0.0 0.0 4.43 6.49 0.63 2.74 0.6 0.0 2.4 0.0
1.0 3.79 0.6 0.58 10.53 4.74 0.0 0.54 0.23 0.0 0.0 0.5 6.93 0.0 0.94 0.0 0.0 0.0 0.0
0.0 0.0 0.0 0.0 0.0 0.0 5.15 1.01 0.0 3.11 6.33 0.73 0.0 0.0 5.41 2.74 3.95 0.0 0.0 0.0
1.56 0.46 1.07 8.77 0.0 0.58 1.59 1.32 0.0 0.0 0.0 0.45 4.97 0.0 1.19 0.0 0.54 0.0
0.51 0.88 0.0 0.0 0.29 1.63 1.64 0.0 2.09 0.0 0.79 0.45 0.31 1.47 5.96 0.0 0.0 0.0 0.0
0.61 0.0 0.86 0.0 0.0 2.87 0.66 0.0 0.0 2.28 1.17 3.9 2.55 0.0 0.0 0.0 0.0 0.2 0.0 1.98
0.39 0.0 0.38 0.0 0.62 0.78 0.93 0.34 0.0 1.39 0.0 0.95 0.0 1.37 0.0 0.0 0.0 1.21 0.0
0.0 0.0 0.0 3.74 0.0 0.0

touch_screen.vals

>percent minority
76.82 73.79 70.91 70.85 96.01 68.13 67.86 66.88 66.85 66.73 66.36 66.1 65.4 64.88
63.78 63.49 62.7 62.43 62.35 62.3 61.62 61.27 60.77 60.59 60.22 59.48 59.29 58.58
57.67 55.91 54.86 54.04 52.58 52.21 51.89 51.74 51.35 48.29 47.68 44.22 43.42 42.42
42.2 42.17 41.57 39.07 36.9 35.85 35.42 34.65 32.05 30.43 29.32 28.97 25.72 24.56
22.25 21.89 18.67 14.32 13.32 12.47
>percent undervote
1.13 0.45 0.32 0.0 1.52 0.23 0.32 1.51 4.0 0.3 1.21 1.98 0.3 0.0 0.16 3.54 1.66 0.0
0.97 0.58 2.94 0.3 0.26 0.3 0.66 0.57 2.82 4.42 0.55 0.91 3.91 0.65 0.18 0.41 0.0 6.47
4.85 0.35 5.56 4.44 5.65 1.97 0.55 7.21 0.0 3.6 3.98 1.32 0.0 0.56 1.37 4.47 3.39 0.83
6.25 9.6 4.53 3.93 3.99 4.47 5.18 4.29

0to25.vals

>push button

822.5 773.0 1181.0 1004.0 715.5 745.0 544.0 729.0 574.0 833.0 788.5 1116.0 1077.5
769.0 715.5 1069.5 740.0 943.5 558.0 961.5 685.0 947.0 832.0 871.0 1103.0 993.5 916.0
990.0 1029.0 794.0 825.0 834.0 863.0 828.0 1088.0 406.5 915.0 691.0 400.0 1009.0 497.5
910.0 1015.0 862.0 701.0 722.0 840.0 876.5 577.0 831.0 636.0 892.0 943.5 655.0 876.5
965.0 460.0 293.0 509.5 1112.0 557.0 987.0 1288.0 265.5 760.0 901.0 814.5 548.0 921.0
1044.0 537.5 232.0 1288.0 533.0 1288.0 439.0 246.0 38.0 970.0 1108.0 1174.0 1192.0
1155.5 1176.0 1203.5 1075.0 1189.0 1199.0 965.0 31.0 822.5 278.0 214.0 998.5 482.0
839.0 882.0 248.0 367.0 1159.0 1048.0 889.0 1076.0 1145.5 729.0 1092.5 1096.0 1100.0
956.0 1036.0 971.0 1169.0 718.0 881.0 187.0 653.0 297.0 885.0 927.0 1005.0 1288.0
869.5 902.5 1069.5 1132.0 1052.5 537.5 1160.0 977.0 228.0 937.0 580.5 982.0 947.0
912.0 481.0 899.5 508.0 1052.5 878.5 634.0 822.5 704.5 109.0 222.5 356.5

>touch screen

108.0 712.0 483.5 647.0 605.0 593.0 699.0

>optical scan

1097.5 1137.0 1288.0 1288.0 516.0 1007.0 985.0 897.5 1059.0 723.0 752.0 1022.5 939.0
1288.0 1288.0 1113.0 1170.0 1144.0 1288.0 1288.0 1288.0 1288.0 1288.0 1214.0 1288.0
1288.0 1087.0 1288.0 1288.0 1288.0 1057.0 1288.0 1196.5 1288.0 1143.0 1288.0 1209.0
1288.0 1202.0 1288.0 1288.0 1198.0 1110.0 1288.0 1288.0 1288.0 1288.0 1205.0 1150.0
1227.0 1288.0 1288.0 1153.5 1288.0 1115.0 1288.0 1288.0 1135.0 1288.0 1288.0 1288.0
1191.0 1221.0 1163.5 1171.0 1288.0 1185.0 1288.0 1288.0 1090.0 1288.0

25to50.vals

>push button

803.5 1035.0 920.0 872.5 531.0 864.5 955.0 979.0 1002.0 1106.0 779.0 936.0 895.5 925.0
803.5 1288.0 919.0 488.0 843.5 1058.0 1012.0 989.0 1166.0 464.5 899.5 818.5 906.5
463.0 244.0 818.5 930.5 35.0 1033.5 1028.0 1003.0 1038.0 745.0 651.5 872.5 745.0
1288.0 1065.5 830.0 983.5 670.0 926.0 1085.0 135.0 380.5 718.0 392.0 660.0 1010.0
294.5 502.0 859.5 751.0 796.0 447.5 814.5 673.0 566.0 439.0 372.5 628.5 543.0 401.0
470.0 986.0 849.5 893.5 411.0 941.5 663.0 637.5 1001.0 350.0 676.0 836.0 420.0 175.0
649.0 756.5 774.0 743.0 969.0 749.5 589.0 690.0 398.0 209.0 759.0 394.5 453.5 1065.5
754.0 1071.5 666.5 529.0 466.5 851.0 614.0 957.0 729.0 625.0 810.0 1022.5 784.0 496.0
365.0 1027.0 930.5 808.5 619.5 445.0 1015.0 880.0 1015.0 188.0 471.0 664.0 992.0 695.0
494.0 713.0 564.5 1161.5 951.0 736.0 803.5 1073.0 637.5 507.0 822.5 288.0 623.0 779.0
914.0 835.0 688.0 867.0 406.5 783.0 788.5 811.0 897.5 546.0 628.5 858.0 786.0 509.5
575.0 792.0 1101.5 642.0 856.5 560.5 549.0 678.0 998.5 895.5 375.0 613.0 580.5 462.0
616.0 308.0 596.5 567.0 428.5 599.5 675.0 852.0 843.5 849.5 973.0 739.0 603.0 793.0
127.0 23.0 960.0 385.0 473.0 1081.0 478.0 164.0 596.5 725.0 634.0 909.0 1095.0 563.0
1101.5 930.5 911.0 356.5 569.0 945.0 1041.0 888.0 1037.0 980.0 949.5 845.5 729.0 703.0
517.0 1018.0 934.0 977.0 1008.0 869.5 1068.0 316.0 390.0 233.0 495.0 408.5 747.0 570.5
704.5 640.0 112.0 65.0 458.0 1288.0 1288.0 1194.5 1025.5 1186.5 1134.0 1288.0 1151.5
1288.0 1222.0 1157.5 1142.0 1056.0 1117.0 1104.0 1060.0 1201.0 657.0 436.0 820.0 763.0
769.0 583.5 679.0 237.0 377.5 388.0 479.0 226.5 733.0 902.5 262.0 476.5 890.5 627.0
137.0 419.0 578.0 974.0 418.0 753.0 670.0 213.0 476.5 1081.0 908.0 1089.0 1052.5 827.0
1114.0 1125.5 1077.5 1013.0 842.0 993.5 1046.0 1063.0 655.0 1055.0 923.0 1033.5 1083.5
588.0 958.0 692.0 1022.5 935.0 977.0 468.0 485.0 1109.0 837.0 996.0 639.0 866.0 758.0
474.0 493.0 734.0 607.0 702.0 349.0 781.0 363.0 777.0 683.0 1042.5 662.0 697.0 698.0
514.0 965.0 416.0 826.0 769.0 587.0 555.5 874.0 1107.0 856.5 807.0 1030.0 800.0 328.0
661.0 859.5 632.0 579.0 282.0 797.0 269.0 1045.0 651.5 726.0 965.0 582.0 981.0 1019.0
499.5 591.5 928.0 737.0 238.0 922.0 670.0 742.0 761.5 1006.0 779.0 1288.0 886.0 528.0
547.0 247.0 453.5 303.0 451.0 972.0 938.0 1288.0 209.0 226.5 140.0 69.0 610.0 1288.0

>touch screen

1025.5 812.5 244.0 428.5 344.0 604.0 417.0 610.0 700.0 1288.0 771.5 1180.0 1288.0
1091.0 1148.0 1097.5 1208.0 1182.0
>optical scan
1193.0 1288.0 917.5 855.0 487.0 433.0 313.5 1155.5 709.5 273.0 423.0 525.0 287.0 615.0
335.0 741.0 961.5 447.5 553.0 917.5 812.5 1062.0 1288.0 1288.0 1288.0 1140.5 1288.0
1288.0 1288.0 1288.0 1288.0 1207.0 149.0 274.0 1288.0 1288.0 1212.0 1184.0 1167.5
1119.0 1288.0 1288.0 1288.0 1031.5 1288.0 1288.0 1188.0 1129.0 1288.0 1049.0 1288.0
1288.0 1178.0 1288.0 1186.5 1065.5 1288.0 1194.5 1288.0 1225.0 1288.0 1175.0 1288.0
1172.5 1203.5 1288.0 1288.0 1288.0 1288.0 1288.0 1167.5 1288.0 1288.0 1172.5 1288.0
1125.5 1136.0 72.5 1288.0

50to75.vals

>push button
191.0 1099.0 576.0 771.5 598.0 1052.5 1.0 359.0 648.0 940.0 803.5 847.0 845.5 767.0
650.0 464.5 1086.0 594.0 537.5 339.5 1124.0 241.0 322.0 550.0 321.0 714.0 195.0 775.0
486.0 749.5 480.0 377.5 522.0 841.0 787.0 904.0 520.0 435.0 304.0 342.0 206.0 456.0
346.0 983.5 504.0 387.0 261.0 354.0 503.0 552.0 257.0 323.0 954.0 619.5 541.0 310.0
404.0 623.0 424.5 518.0 84.0 254.0 444.0 953.0 358.0 764.5 595.0 447.5 344.0 798.5
260.0 735.0 399.0 527.0 764.5 313.5 410.0 499.5 505.5 324.0 347.0 949.5 309.0 167.0
88.0 133.0 182.0 305.0 117.0 838.0 1127.5 62.0 70.0 332.0 60.0 367.0 334.0 47.0 372.5
501.0 136.0 626.0 853.0 829.0 348.0 369.0 1020.0 447.5 307.0 497.5 803.5 707.0 718.0
782.0 331.0 283.5 455.0 329.0 512.5 687.0 884.0 968.0 941.5 1017.0 975.0 817.0 933.0
610.0 523.0 998.5 199.5 617.0 756.5 562.0 591.5 890.5 1031.5 641.0 803.5 875.0 668.0
515.0 680.5 666.5 120.0 231.0 505.5 92.0 330.0 234.5 791.0 1288.0 1288.0 1083.5 1288.0
1094.0 1127.5 1130.0 1123.0 965.0 1288.0 1047.0 255.0 294.5 610.0 194.0 240.0 677.0
755.0 816.0 102.0 988.0 20.0 524.0 674.0 258.0 83.0 360.0 222.5 1145.5 848.0 71.0
706.0 489.0 147.5 1288.0 708.0 905.0 684.0 1092.5 259.0 868.0 959.0 878.5 721.0 160.0
333.0 56.0 143.0 2.0 21.0 156.0 185.0 519.0 530.0 43.0 732.0 1039.0 320.0 542.0 391.0
795.0 85.0 798.5 103.0 924.0 606.0 808.5 602.0 352.0 265.5 490.0 686.0 126.0 337.0
526.0 766.0 570.5 225.0 947.0 913.0 216.0 601.0 655.0 111.0 48.0 585.0 412.5 113.0
229.5 114.0 138.0 268.0 61.0 155.0 428.5 263.0 205.0 41.0 171.0 10.0 403.0
>touch screen
317.0 720.0 540.0 906.5 785.0 695.0 621.0 883.0 1163.5 1211.0 1050.0 1183.0 1228.0
1218.0 1220.0 1165.0 1224.0 1133.0 1140.5 1196.5 1179.0 1288.0 1022.5 1215.0 1074.0
1288.0 1213.0 1226.0 1229.0 1111.0 1177.0 1288.0 1216.0 1200.0 1288.0
>optical scan
220.0 572.5 952.0 535.0 234.5 893.5 534.0 355.0 131.5 644.5 537.5 198.0 201.0 380.5
689.0 396.5 776.0 1288.0 1153.5 1288.0 1147.0 1288.0 1288.0 1121.0 405.0 1157.5 1206.0
1288.0 1210.0 1288.0 1288.0 1288.0 1219.0 1139.0 1288.0 1288.0 1149.0 1288.0 1040.0
1288.0 1288.0 1288.0 1288.0 1122.0 1061.0

75to100.vals

>push button
377.5 590.0 253.0 319.0 45.0 748.0 729.0 371.0 568.0 586.0 458.0 272.0 142.0 370.0
172.0 428.5 545.0 532.0 472.0 236.0 432.0 298.5 377.5 437.0 105.0 224.0 209.0 184.0
154.0 134.0 173.0 341.0 394.5 469.0 110.0 434.0 267.0 280.0 26.0 443.0 13.0 99.0 52.0
122.0 8.0 491.5 75.0 158.0 511.0 630.0 147.5 461.0 339.5 724.0 157.0 583.5 412.5 290.0
179.0 219.0 167.0 466.5 86.0 325.0 386.0 139.0 176.0 555.5 1011.0 97.0 82.0 458.0
106.5 15.0 33.0 57.5 78.5 4.0 54.0 1065.5 363.0 150.5 296.0 256.0 174.0 55.0 229.5
658.0 22.0 631.0 218.0 300.0 276.0 59.0 53.0 665.0 512.5 344.0 93.0 124.5 402.0 87.0
3.0 141.0 682.0 301.0 283.5 367.0 212.0 1131.0 450.0 711.0 1288.0 6.5 57.5 277.0 130.0
180.0 264.0 244.0 336.0 242.0 302.0 693.0 442.0 995.0 1105.0 1079.0 393.0 131.5 279.0
153.0 25.0 30.0 560.5 161.0 211.0 186.0 152.0 159.0 167.0 118.5 998.5 384.0 306.0
312.0 39.0 170.0 196.0 169.0 559.0 12.0 16.0 6.5 5.0 63.0 189.0 193.0 298.5 32.0 106.5

353.0 190.0 124.5 291.0 67.0 144.0 96.0 239.0 118.5 66.0 217.0 181.0 163.0 285.0 415.0
521.0 554.0 311.0 424.5 672.0 315.0 864.5 338.0 738.0 634.0 422.0 115.0 930.5 215.0
1042.5 249.0 414.0 197.0 1288.0 680.5 643.0 483.5 68.0 761.5 183.0 452.0 790.0 203.0
271.0 199.5 361.0 991.0 431.0 192.0 98.0 286.0 695.0 610.0 100.5 363.0 491.5 78.5
854.0 439.0 551.0 426.0 289.0 202.0 709.5 165.0 204.0 146.0 44.0 77.0 72.5 81.0 178.0
861.0 14.0 49.0 50.0 17.0 64.0 177.0 29.0 76.0 129.0 221.0 123.0 251.0 599.5 275.0
128.0 94.0 104.0 95.0 74.0 421.0 34.0 207.0 162.0 28.0 91.0 36.0 292.0 145.0 281.0
251.0 572.5 408.5 19.0 623.0 659.0 887.0 382.5 270.0 326.5 475.0 396.5 326.5 441.0
374.0 351.0 564.5 618.0 150.5 646.0 51.0 389.0 11.0 24.0 42.0 27.0 18.0 89.0 121.0
37.0 40.0 80.0 100.5 251.0 116.0 46.0 9.0
>touch screen
1071.5 1118.0
>optical scan
90.0 382.5 644.5 318.0 1151.5 1288.0 1288.0 1223.0 1120.0 1288.0 1288.0 1190.0 1288.0
1288.0 1161.5 1081.0 1138.0 1288.0 1217.0

Printed in the United States
by Baker & Taylor Publisher Services